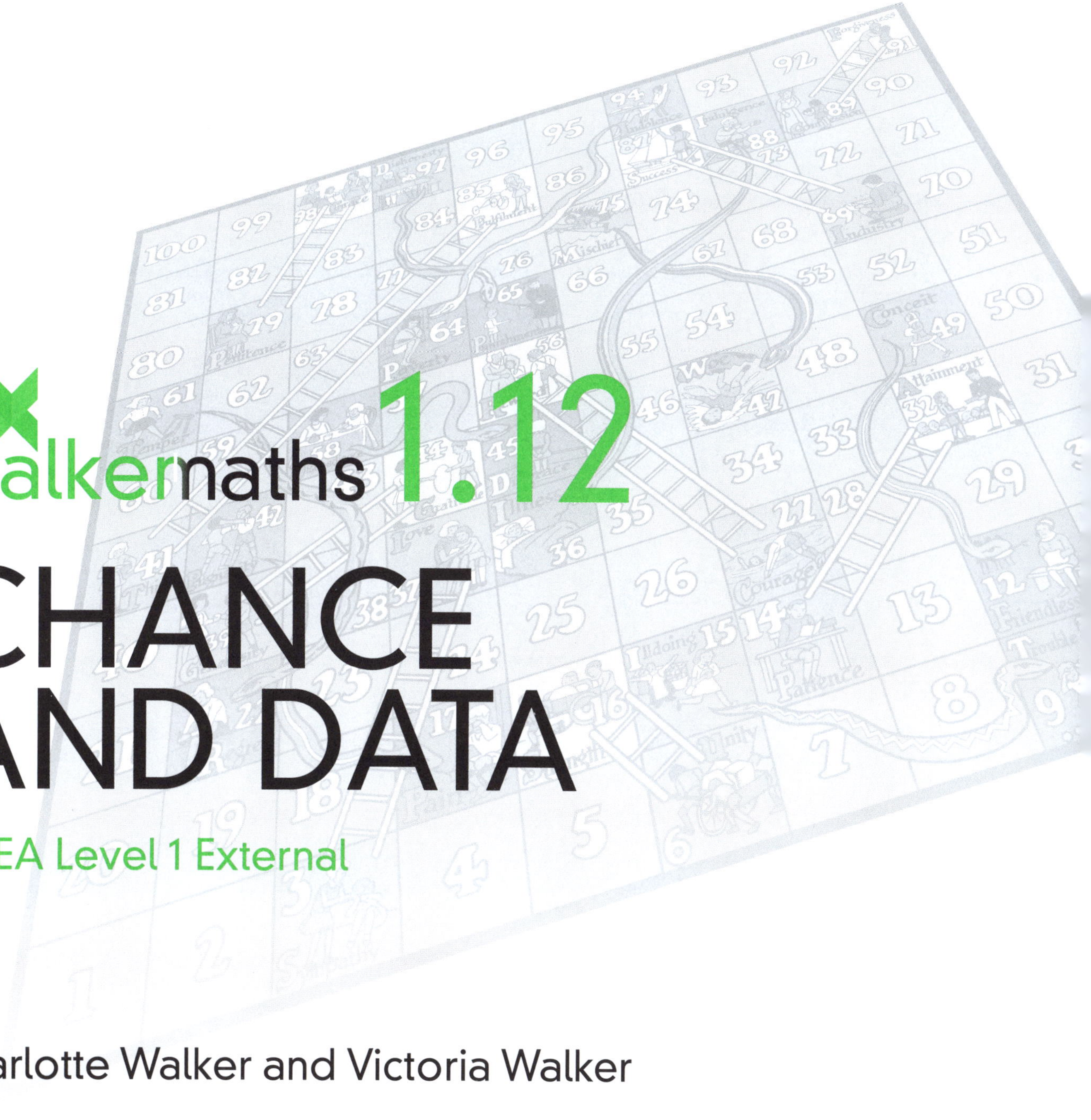

walkermaths 1.12

CHANCE AND DATA

NCEA Level 1 External

Charlotte Walker and Victoria Walker

Walker Maths 1.12 Chance and Data
1st Edition
Charlotte Walker
Victoria Walker

Editor: Eva Chan
Designer: Cheryl Smith, Macarn Design
Production controller: Siew Han Ong

Any URLs contained in this publication were checked for currency during the production process. Note, however, that the publisher cannot vouch for the ongoing currency of URLs.

Acknowledgements
Cover photo courtesy of Shutterstock.

For product information and technology assistance,
in Australia call **1300 790 853**;
in New Zealand call **0800 449 725**

For permission to use material from this text or product, please email
aust.permissions@cengage.com

National Library of New Zealand Cataloguing-in-Publication Data
A catalogue record for this book is available from the National Library of New Zealand.

978 0 17 037042 4

Cengage Learning Australia
Level 7, 80 Dorcas Street
South Melbourne, Victoria Australia 3205

Cengage Learning New Zealand
Unit 4B Rosedale Office Park
331 Rosedale Road, Albany, North Shore 0632, NZ

For learning solutions, visit **cengage.co.nz**

Printed in China by 1010 Printing International Limited.
15 16 17 25 24 23

CONTENTS

Glossary

Make your own glossary of key terms:

Term	Definition	Picture/Example
Data		
Minimum		
Lower quartile		
Median		
Mode		
Upper quartile		
Maximum		
Inter-quartile range (IQR)		
Range		
Mean		

ISBN: 9780170370424

Term	Definition	Picture/Example
Sample		
Population		
Frequency		
Probability		
Proportion		
Percentage		
Either		
Experimental probability		
Theoretical probability		
Trend		
Cycle		
A fair die		

ISBN: 9780170370424

Statistical enquiry cycle

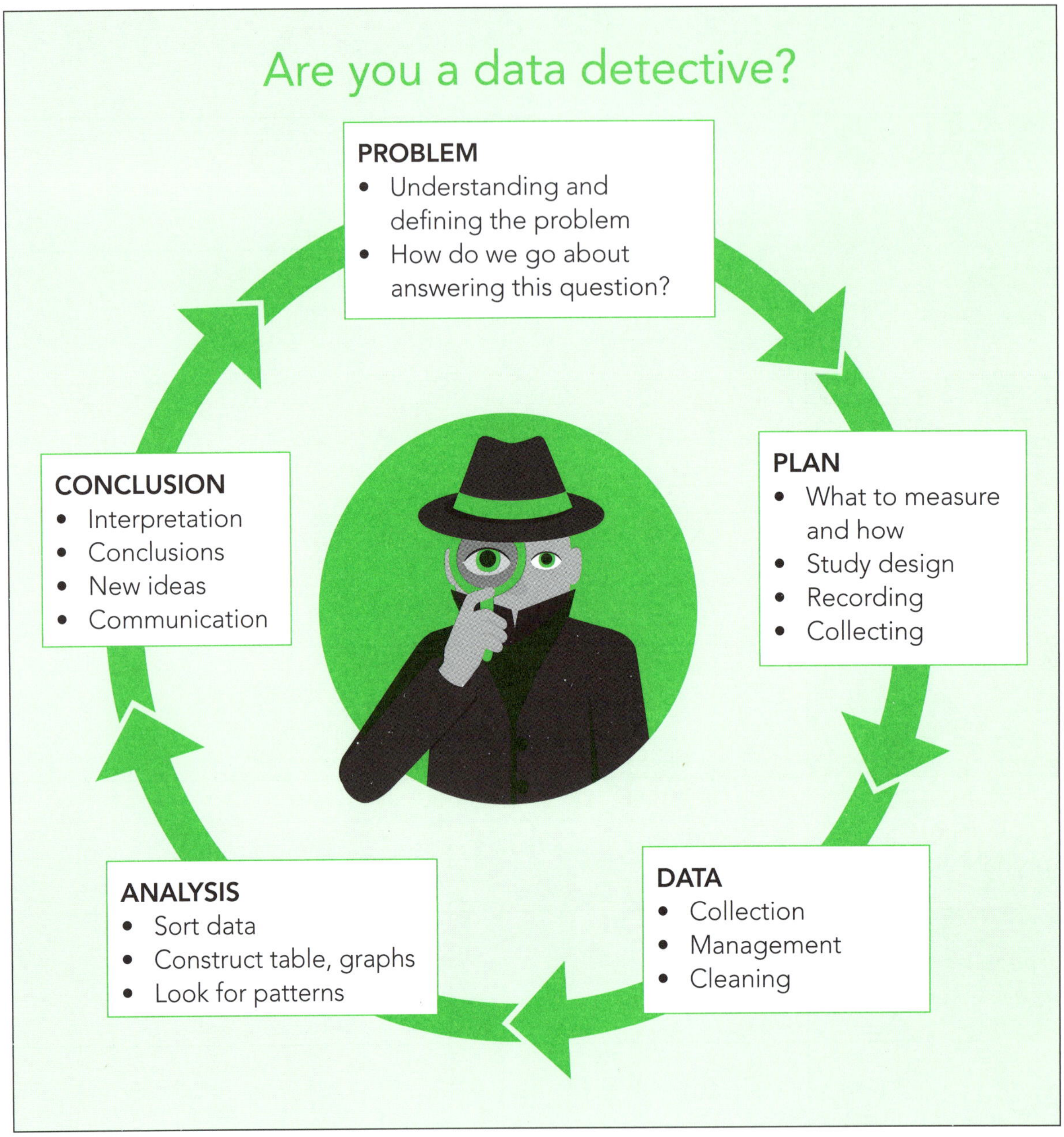

Are you a data detective?

PROBLEM
- Understanding and defining the problem
- How do we go about answering this question?

PLAN
- What to measure and how
- Study design
- Recording
- Collecting

DATA
- Collection
- Management
- Cleaning

ANALYSIS
- Sort data
- Construct table, graphs
- Look for patterns

CONCLUSION
- Interpretation
- Conclusions
- New ideas
- Communication

ISBN: 9780170370424

Probability

The range of values for probabilities

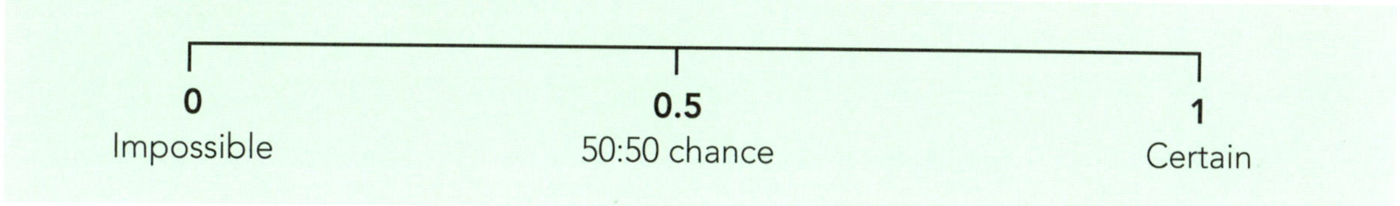

Discuss the meanings of the following words and phrases with others in your class, and match each to its appropriate probability value(s). You may have several words beside the same probability value, and some words can cover several probability values.

Very unlikely, improbable, slight chance, a sure thing, extremely likely, maybe, likely, impossible, possible, probable, unlikely, no chance, definite, no way

0	
0.1	
0.2	
0.3	
0.4	
0.5	
0.6	
0.7	
0.8	
0.9	
1.0	

ISBN: 9780170370424

Using numbers for writing probability

- Probabilities can be written as **fractions, decimals or percentages**.
- You can convert between these with your calculator.
- When you want to compare probabilities, you should always use **decimals**.

Example: Write the probability $\frac{7}{16}$ as a decimal. 7 [÷] 16 [=] 0.4375

Convert the following probabilities to decimals, and state which of each pair is more likely.

1 $\frac{7}{16}$ = ________ $\frac{1}{3}$ = ________

Most likely: ________________

2 $\frac{3}{4}$ = ________ $\frac{8}{11}$ = ________

Most likely: ________________

3 $\frac{17}{18}$ = ________ $\frac{19}{20}$ = ________

Most likely: ________________

4 $\frac{13}{24}$ = ________ $\frac{27}{50}$ = ________

Most likely: ________________

5 $\frac{11}{20}$ = ________ $\frac{7}{13}$ = ________

Most likely: ________________

6 $\frac{6}{17}$ = ________ $\frac{3}{8}$ = ________

Most likely: ________________

Test yourself

For the following situations, indicate whether each is right or wrong, and justify your decision.

1 Nikki said that the probability is 0.009 that tomorrow will be a sunny day.

____

2 Adam calculated that the probability was -0.35 that he would win a game.

____

3 Tahu calculated that the probability that he would pass Level 1 was 105%.

____

4 Sora said that the probability was $\frac{1}{200}$ that she would be late for school.

____

 ISBN: 9780170370424

Ways of calculating probabilities

1 Equally likely outcomes

$$\text{Probability} = \frac{\text{number of favourable outcomes}}{\text{total possible outcomes}}$$

Example 1: For a fair die, P(1 or a 2) = $\frac{2}{6}$ or $\frac{1}{3}$ or $0.\dot{3}$

Example 2: Using a pack of cards with no jokers (52 cards):

P(spade) = $\frac{13}{52}$ or $\frac{1}{4}$ or 0.25

P(ace of spades) = $\frac{1}{52}$ or 0.01923

P(ace) = $\frac{4}{52}$ or $\frac{1}{13}$ or 0.0769

P(spade or ace) = $\frac{16}{52}$ or 0.1154

Calculate the following probabilities.
When tossing a fair die:

1 P(6) = ________

2 P(4, 5 or 6) = ________

3 P(odd number) = ________

4 P(not a 6) = ________

5 P(7) = ________

6 P(1, 2, 3, 4, 5 or 6) = ________

Using a pack of cards with no jokers (52 cards), and selecting one card at random:

7 P(club) = ________

8 P(black card) = ________

9 P(10) = ________

10 P(black 10) = ________

11 P(5 or 6) = ________

12 P(red 5, red 6 or red 7) = ________

13 P(not an ace) = ________

14 P(not a king or queen) = ________

Mike has a bag of lollies. It contains 7 red lollies, 3 green ones, 5 yellow ones and 9 white ones. He puts his hand in the bag and selects a lolly at random:

15 P(red) = ________

16 P(not getting a red) = ________

17 P(red or green) = ________

18 P(not getting a red or green) = ________

19 P(white, red or green) = ________

20 P(not yellow) = ________

21 P(black) = ________

22 P(red, green, yellow or white) = ________

ISBN: 9780170370424

Frank is playing a board game in which two dice are tossed at the same time. One is black and the other is green. The possible outcomes are shown in the table below.

	1	2	3	4	5	6
1	1 1	1 2	1 3	1 4	1 5	1 6
2	2 1	2 2	2 3	2 4	2 5	2 6
3	3 1	3 2	3 3	3 4	3 5	3 6
4	4 1	4 2	4 3	4 4	4 5	4 6
5	5 1	5 2	5 3	5 4	5 5	5 6
6	6 1	6 2	6 3	6 4	6 5	6 6

23 How many different possible outcomes are there? ______________________

24 List the outcomes that produce a total of 3. ______________________

25 P(total is 3) = ______________

26 P(total is 5) = ______________

27 P(total is 3 or 5) = ______________

28 P(total is more than 10) = ______________

29 P(total is 10 or less) = ______________

30 P(total is even) = ______________

31 P(double) = ______________

32 P(green die is more than black die) = ______

In another game, Frank uses a spinner with the numbers 1, 2, 4 and 8, and he tosses a die.

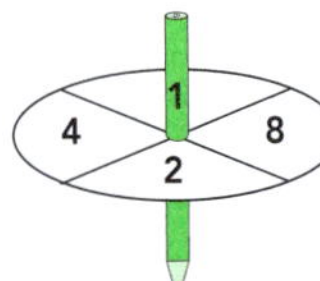

33 Use the space below to draw a table of the possible outcomes.

34 P(total is 2) = ______________

35 P(double) = ______________

36 P(odd) = ______________

37 P(total is more than 10) = ______________

 ISBN: 9780170370424

2 Long run relative frequency

- Sometimes the probability of an event is **difficult or impossible** to calculate.
- In these cases we do **many trials** and record the number of times an event occurs.
- Usually the **true value** of the probability will **never be known**.
- **The greater the number of trials, the closer our estimate for the probability will be to the true probability.**

$$\textbf{Probability} = \frac{\textbf{number of favourable outcomes}}{\textbf{total possible outcomes}}$$

Example: Hugo kept a record of how he travelled to school last year. He went to school on 178 days. He biked on 102 days, walked on 24 days, and caught the bus on the other days. Calculate the probabilities that he biked, walked and caught the bus.

P(Hugo **bikes** to school) $= \frac{\text{number of times he has biked to school in the last year}}{\text{total number of school days in the last year}}$

$= \frac{102}{178} = 0.5730$

P(Hugo **walks** to school) $= \frac{\text{number of times he has walked to school in the last year}}{\text{total number of school days in the last year}}$

$= \frac{24}{178} = 0.1348$

P(Hugo **buses** to school) $= \frac{\text{number of times he has bused to school in the last year}}{\text{total number of school days in the last year}}$

$= \frac{52}{178} = 0.2921$

Answer the following questions.

1 Marama was in the same class as Hugo, and she also kept records of how she got to school. She went to school on 180 days. She caught the bus on 126 days, walked on 41 days, and her mum drove her on the remaining days. Calculate the probabilities that she caught the bus, she walked and she was driven.

ISBN: 9780170370424

2 There are 28 students in their class. In their probability test, 7 students got Excellence grades, 13 got Merit, 6 got Achieved grades, and 2 got Not Achieved. Calculate the probabilities of getting each grade.

3 A survey of the time spent watching television each school day by 160 secondary school students produced the following results.

Time	Number of students
None	36
0 < 1 hour	59
between 1 and 2 hours	29
between 2 and 3 hours	15
over 3 hours	21

a What is the probability that a student watches no television?

b What is the probability that a student watches less than one hour of television each school day?

c What percentage of students watch more than one hour of television each school day?

4 The same survey asked the same students about their lunches on school days.

Source of lunch	Number of students
Had no lunch	13
Packed lunch from home	74
Bought on the way to school	25
Bought from canteen	48

a What is the probability that a student has no lunch?

b What is the probability that a student bought their lunch?

c What percentage of students eat lunch?

5 The same survey asked the same students how they had got to school that morning.

How they got to school	Number of students
Walked	39
Scooter or bicycle	17
On a bus	52
Motor bike	9
By car — drove themselves	16
By car — driven by somebody else	27

a How many students walked to school?

b What is the probability that a student came to school on a bus?

c What is the probability that a student came to school by motorised transport?

 ISBN: 9780170370424

6 According to Murphy's law, if you drop a piece of buttered toast, it will land buttered side down. Ngaire wanted to know whether this was true for toast with jam. She conducted an experiment in which she dropped pieces of toast spread with jam, and recorded whether the jam side was 'down' or 'up' when it landed. Her results are summarised in the table below. Round the data to 2 dp.

Trial number	Result	Total 'downs'	% of drops	P ('down')
1	down	1	$\frac{1}{1} = 100$	1.0
2	up	1	$\frac{1}{2} = 50$	0.5
3	down	2	$\frac{2}{3} = 67$	0.67
4	down	3	$\frac{3}{4} = 75$	0.75
5	up	3		
6	down	4		
7	up	4		
8	up	4		
9	up	4		
10	down	5		
11	down	6		
12	down	7		
13	up	7		
14	down	8		
15	up	8		
16	down	9		
17	down	10		
18	up	10		
19	down	11		
20	down	12		

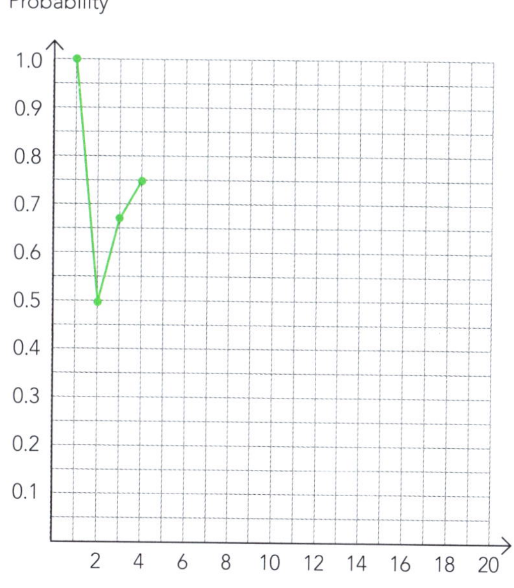

a Complete the table.

b Complete the graph.

c Based on this experiment, what is the probability that the next piece of toast that Ngaire drops will land jam side down?

____

d How certain can she be of this probability?

____

____

ISBN: 9780170370424

Expected number of outcomes

Expected number of outcomes = P(event) x number of trials

Example 1: Amy tosses a die 120 times. How many times can she expect a 6?

Expected number of 6s = P(6) x 120
= $0.1\dot{6} \times 120$
= 20

Example 2: The probability of a person being struck by lightning in the USA is $\frac{1}{700\,000}$ in one year. If in 2015 the population of the USA is 322 584 000, calculate the expected number of people who will be struck by lightning during 2015.

Expected number of people struck by lightning = $\frac{1}{700\,000} \times 322\,584\,000$
= 460 or 461

The exact answer is 460.83...., but this is people, so we must round down or up to the nearest whole person.

Answer the following questions.

1 Archie flips a coin 80 times and records the results. How many 'heads' can he expect?

2 Nick tossed a die 90 times and recorded the results. How many times can he expect to get a 5 or a 6?

3 The probability is 0.1 that the person will be left-handed. If your school roll is 793, how many people would you expect to be left-handed at your school?

4 The probability is 0.08 that the boy will be colour blind. If there are 384 boys at your school, how many would you expect to be colour blind?

5 The probability is 0.001 that an egg will have two yolks. If a farm produces 150 dozen eggs each week, how many would you expect to have two yolks?

6 In the game of poker, each player is dealt a hand of five cards. A full house occurs when three of the cards have the same number, and the other two also have the same number, for example 6, 6, 6, 2, 2. The probability is 0.001441 that a hand contains a full house. If 72 hands are dealt during the course of a game, how many of these would be expected to contain a full house?

 ISBN: 9780170370424

Probabilities from bar graphs and frequency histograms

Example: The graph shows the number of siblings (brothers and sisters) of 50 students.

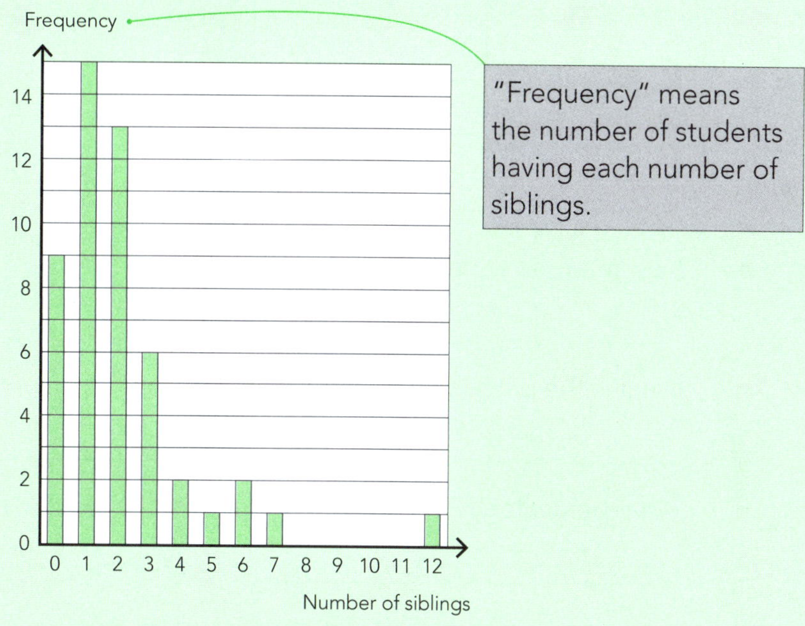

"Frequency" means the number of students having each number of siblings.

a How many students have three siblings?

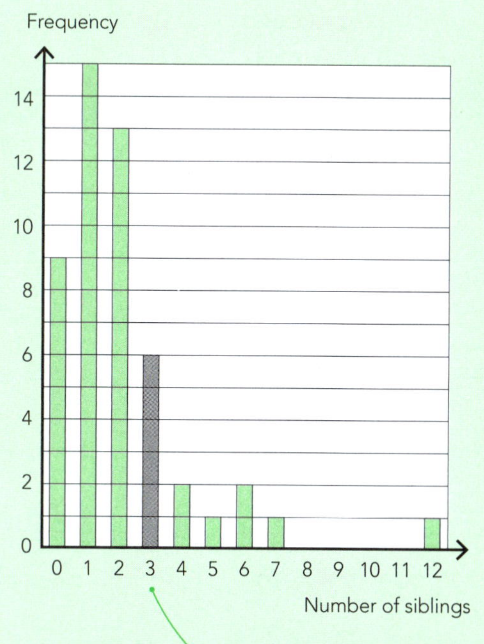

Because the column above 3 goes up to 6, we know that six students have three siblings.

b What is the probability that a student has more than four siblings?

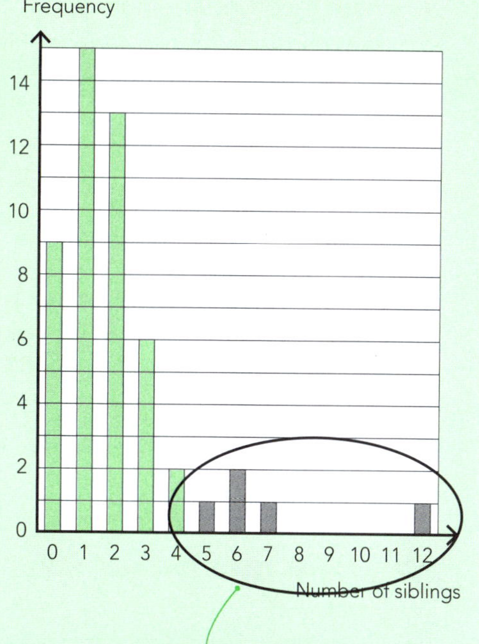

Because there are five students in the columns to the right of 4, and there were 50 students in total, the probability that a student has more than four siblings = $\frac{5}{50}$ = 0.1.

ISBN: 9780170370424

Look at the following graphs and answer the questions.

1 A farmer with a small flock of ewes recorded the number of lambs each animal produced in a season. He plotted these on the graph below.

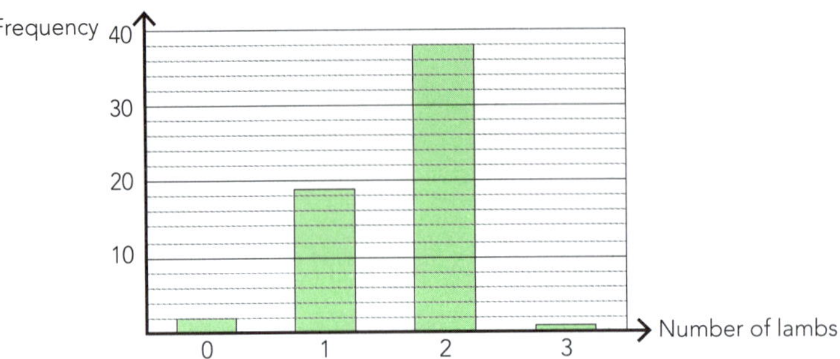

a How many ewes were in his flock?

b What is the probability that a ewe had twins?

c What percentage of ewes had more than one lamb?

d How many lambs in total were produced by the flock?

2 Henare recorded the number of goals scored by his team in each of their hockey games throughout the season. His totals are shown on the graph below.

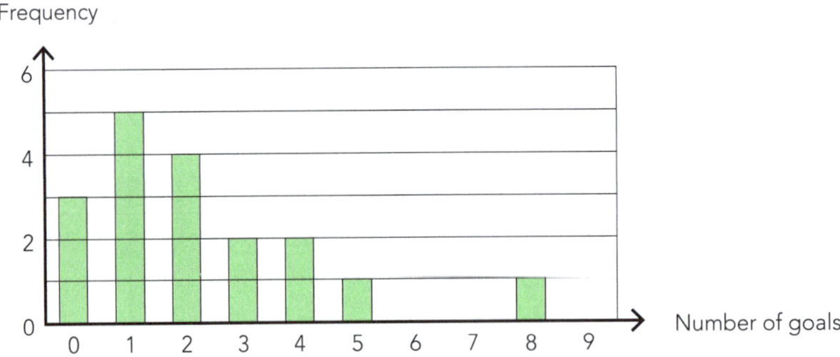

a How many games did they play?

b What is the probability that they scored one goal in a game?

c What is the probability that they scored fewer than three goals in a game?

d How many goals in total did they score during the season?

ISBN: 9780170370424

3 The library recorded how many books people borrowed at any one time. The results are shown in the graph.

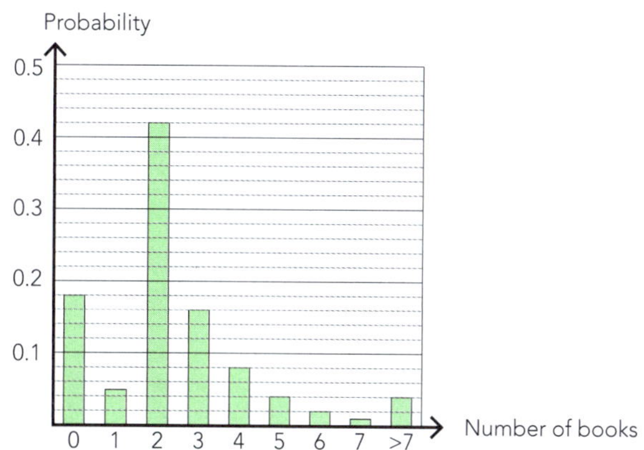

a What is the probability that a person took out no books?

____

b What is the probability that a person took out one book?

____

c What percentage of people took out more than five books?

____

d If there were 15 628 members of the library, how many would be expected to have four books out at any one time?

____

4 A survey was done on how many motor vehicles were owned by the occupants of each house in a small town. The results are shown in the graph.

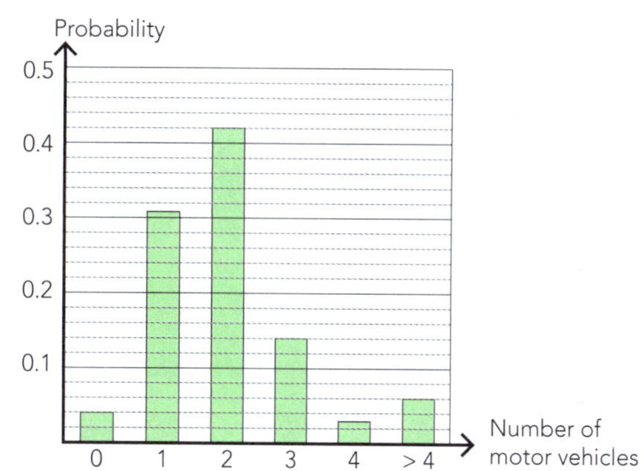

a What is the probability that nobody in a household owned a motor vehicle?

____

b What is the probability that the occupants in a household owned one or two motor vehicles?

____

c What is the probability that the occupants in a household owned more than three motor vehicles?

____

d If there were 986 households in the town, how many would be expected to have three vehicles?

____

ISBN: 9780170370424

5 Ellie performed an experiment by tossing a six-sided die a large number of times and recording the results. The graph of the distribution is shown below.

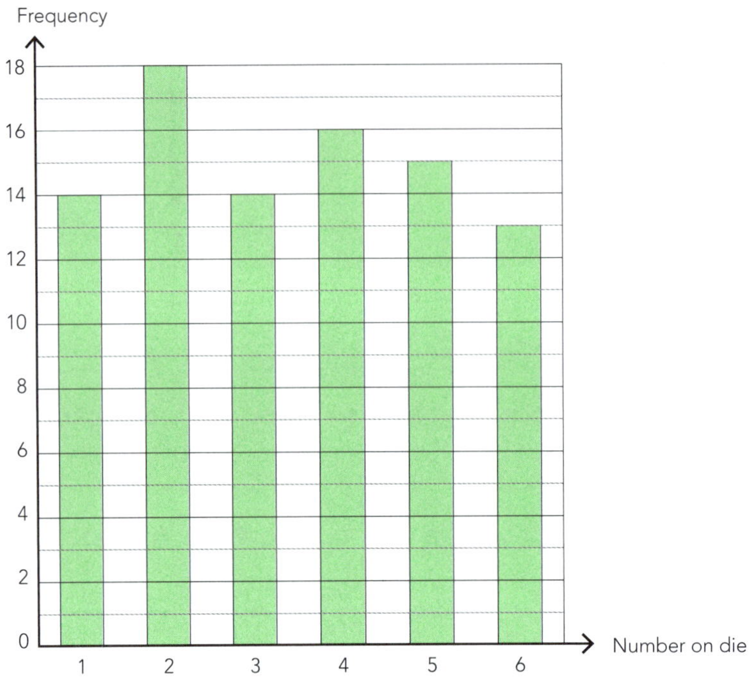

a How many times in total did she toss the die?

b Based on her results, what was her experimental probability of getting a 6?

c What is the theoretical probability of getting a 6 when a fair die is tossed?

d Explain why your answers in **b** and **c** might be different.

e If she tossed a fair die 120 times, how many sixes would you expect her to get?

f If she performed her experiment again, what would you expect her graph to look like?

g She set her computer to do a simulation for 1000 tosses of a die. How would you expect the graph of simulated results to differ from the graph of her experiment?

ISBN: 9780170370424

Combining probabilities — probability trees

Probability trees are very useful for calculating probabilities where several events occur.

Example: Angus had a bag of lollies. It contained 6 red ones and 4 green ones. Without looking, he selected a lolly and ate it. Then he selected a second lolly and ate it.

a What is the probability that he ate two red ones?

b What is the probability that he ate one of each colour?

Steps:

1 Decide what the **events** are, and their order:
Event 1 will be the colour of the first lolly.
Event 2 will be the colour of the second lolly.
Write the **events** at the **ends** of the branches:

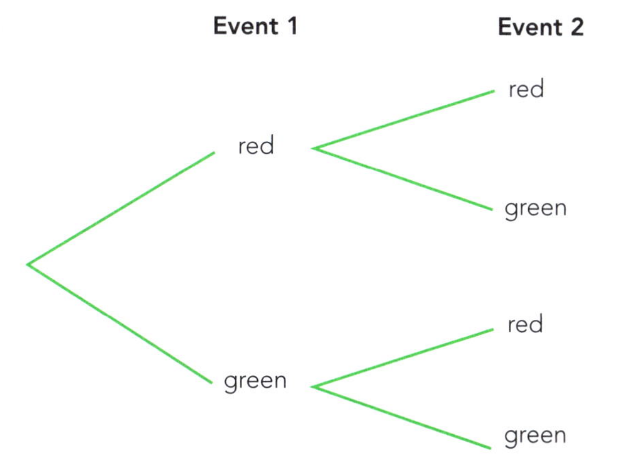

2 Add the **probabilities** of each event to the **middle** of each branch:

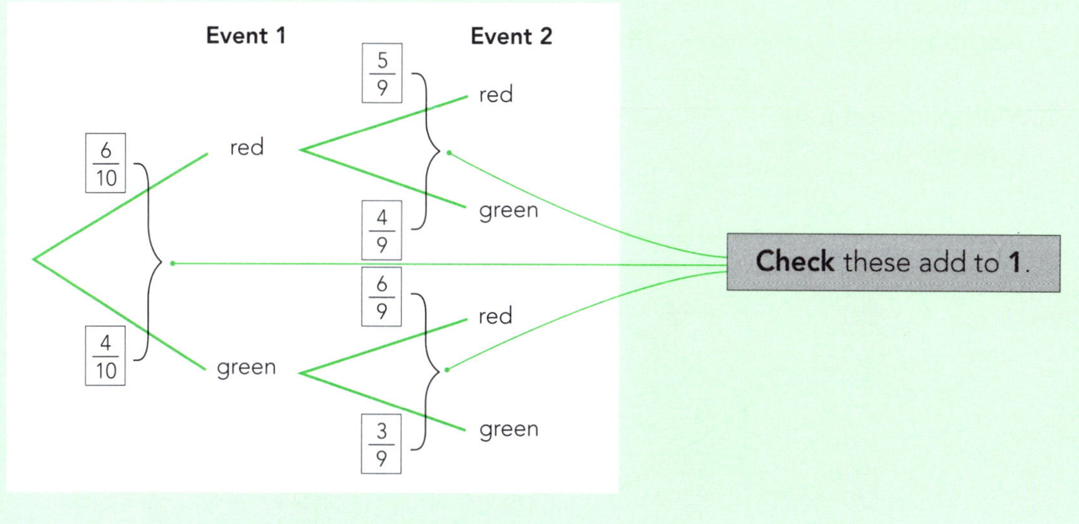

ISBN: 9780170370424

3 **Check** that the probabilities for every branch add to **1**.

4 **List** the outcomes at the ends of each branch, and calculate the probabilities at each end. You **multiply** the probabilities along each branch because Event 1 **and** Event 2 must occur.

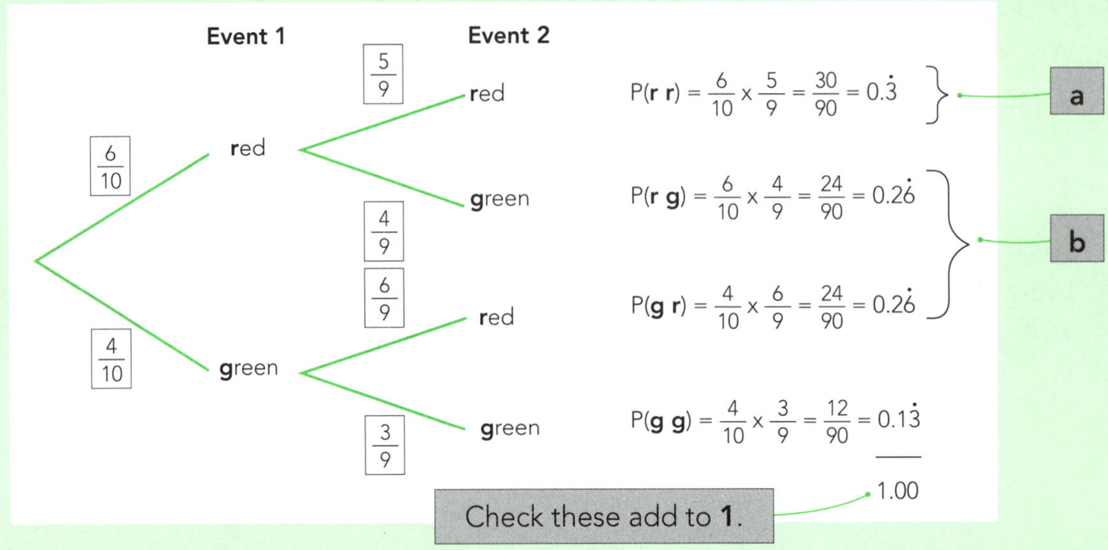

5 **Check** that your probabilities in the right-hand column **add** to **1**. This is because **one** of these options **must** occur.

6 In the two right-hand columns, highlight the event(s) required, along with their probabilities. **Add** these to find the overall probability required.

a $P(r\,r) = 0.\dot{3}$

b $P(r\,g \text{ or } g\,r) = 0.2\dot{6} + 0.2\dot{6} = 0.5\dot{3}$

Useful tips:

1 **Addition Rule:** if one event **OR** another happens ⇒ **ADD** the probabilities.

2 **Multiplication Rule:** if one event **AND** another happens ⇒ **MULTIPLY** the probabilities.

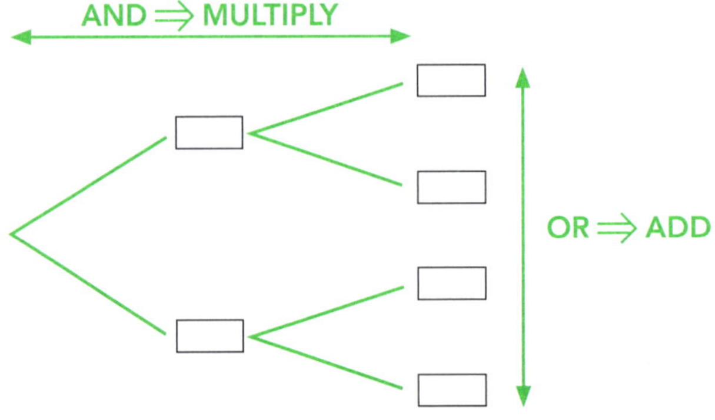

 ISBN: 9780170370424

Use probability trees to answer the following questions.

1 Chris had a bag of lollies. He had 5 yellow ones and 3 blue ones. Without looking, he selected a lolly and then put it back in the bag. He then selected a second lolly.

a Complete the probability tree.

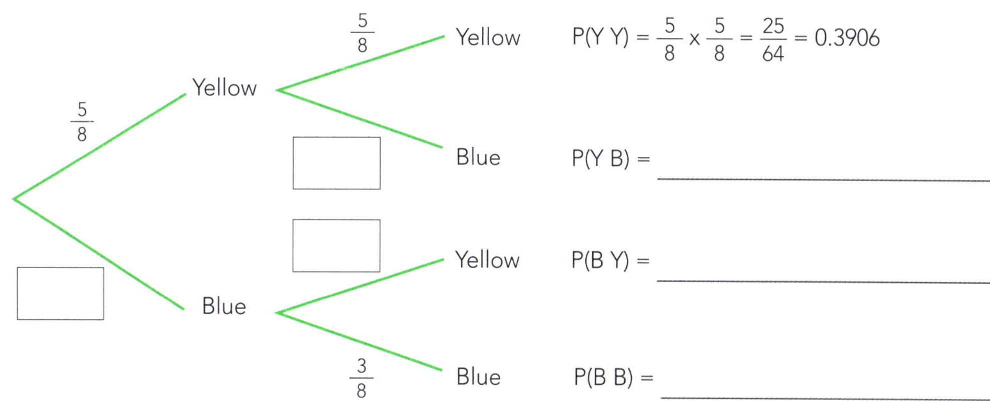

b What is the probability that he selected two yellow ones?

c What is the probability that he selected one of each colour?

d What is the probability that he selected two of the same colour?

e What is the probability that he doesn't select a yellow lolly?

f What is the probability that he selects at least one blue lolly?

g If he repeated this experiment 200 times (without eating any), how many times could he expect to get two yellow lollies?

h If he repeated this experiment 200 times (without eating any), how many times could he expect to get at least one blue lolly?

ISBN: 9780170370424

2 Charlie had a bag of lollies. He had 5 yellow ones and 3 blue ones. Without looking, he selected a lolly and ate it. He selected and ate a second lolly as well.

a Complete the probability tree.

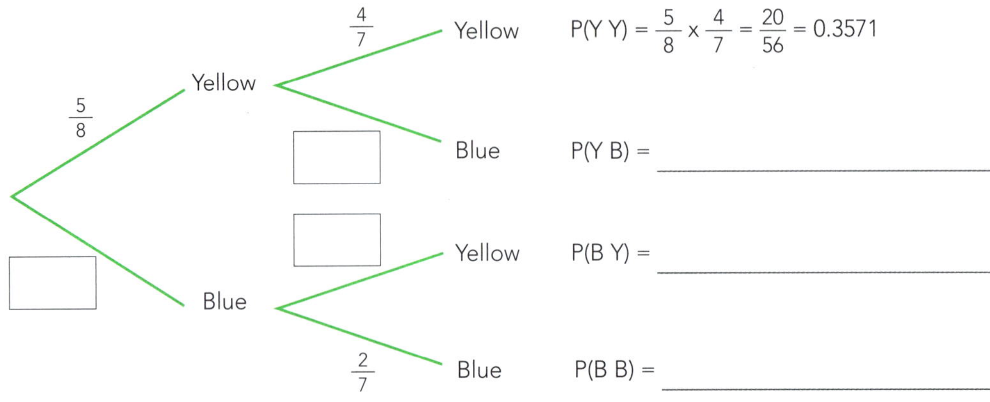

b What is the probability that he ate two yellow ones?

___

c What is the probability that he ate one of each colour?

___

d What is the probability that he ate two of the same colour?

___

e What is the probability that he doesn't eat a yellow lolly?

___

f What is the probability that he ate at least one blue lolly?

___

g What is the probability that after eating his two lollies, there are exactly two blue lollies left in the bag?

___

h What is the probability that after eating his two lollies, there are at least three yellow lollies left in the bag?

___

ISBN: 9780170370424

3 Sarah and three friends are planning their Saturday evening. Two of them want to have pizza and two want hamburgers. Three of them want to go to a movie, and one wants to watch videos. If they all have an equal say and all do the same things, answer the following:

a Complete the probability tree.

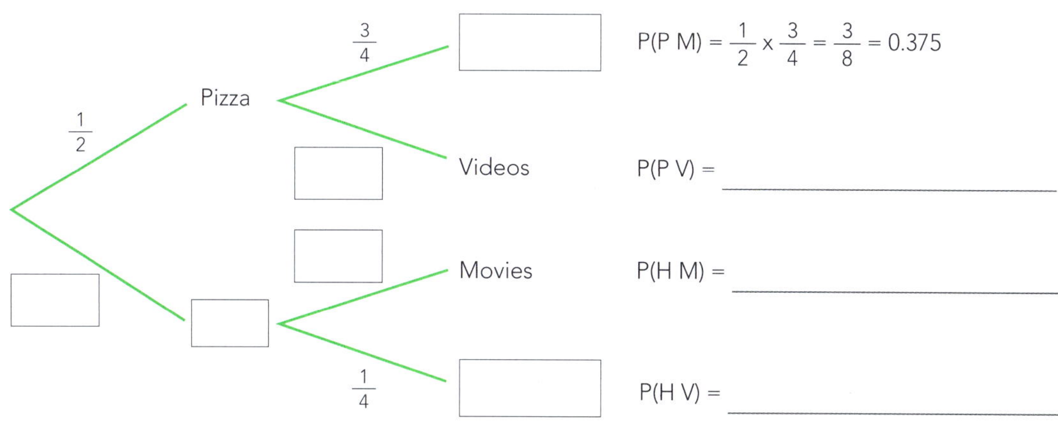

b What is the probability that they go to a movie?

____

c What is the probability that they eat pizza and watch videos?

____

d What is the probability that they eat hamburgers and go to a movie?

____

e What is the probability that they either have pizza or go to a movie?

____

f What is the probability that they either have hamburgers or watch a video?

____

g What is the probability that they neither eat hamburgers nor watch videos?

____

h What is the probability that they neither eat hamburgers nor go to a movie?

____

ISBN: 9780170370424

4 Sometimes when people play Snakes and Ladders, they use the rule that nobody can start the game until they have thrown a six. The probability tree below shows the possible outcomes for the first two throws of the die.

a Complete the probability tree.

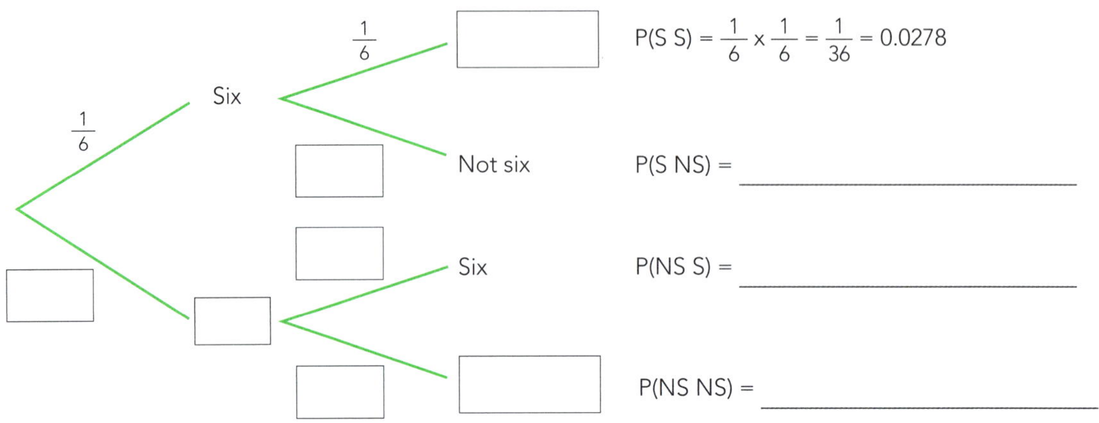

b What is the probability that a player gets a six on the first throw?

______

c What is the probability that a player gets a 'not six' and then a six?

______

d What is the probability that a player gets at least one six in the first two throws?

______

e What is the probability that a player doesn't get a six in the first two throws?

______

f What do you notice about your last two answers? Why is this?

______

g If Thomas recorded the dice throws for the starts of 80 games of Snakes and Ladders, how many would be expected to result in a start within the first two throws?

______

h What is the probability that a player gets to start on the third throw?

______

 ISBN: 9780170370424

5 Georgia devises a game for the school fair. Players contribute a $1 coin and a $2 coin, which they toss. The rules are:

tossing 2 heads ⇒ their $3 is returned, plus an additional $3
tossing 1 head ⇒ they get just their $1 coin back
tossing 0 heads ⇒ they get nothing back.

Complete your own probability tree for this situation.

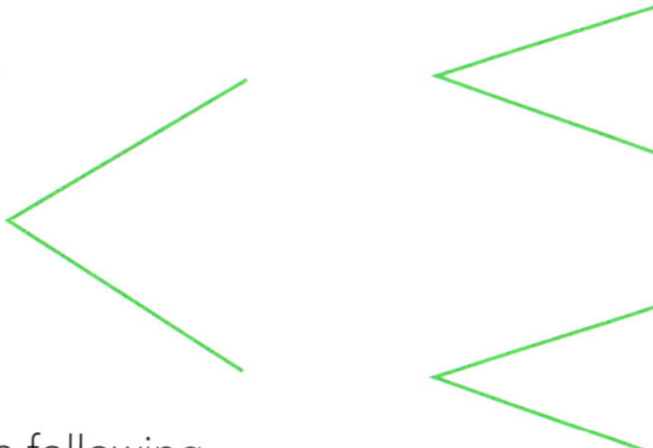

Use it to calculate the following.

a The probability that the $3 is returned, plus an additional $3. ____________

b The probability that a player gets back their $1 coin only. ____________

c What percentage of players get back $1 or less? ____________

d If she repeated this 100 times, how many times would she expect to get no heads?

____

6 A recent survey showed that 80% of students use Facebook and 30% use Twitter. Complete your own probability tree for this situation.

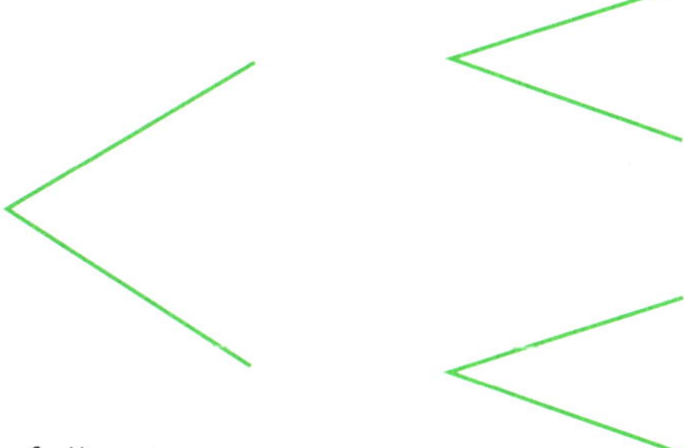

Use it to calculate the following.

a The probability that a student uses both Facebook and Twitter. ____________

b What percentage of students uses just one of the services? ____________

c In a school with 822 students, how many would be expected to use Twitter?

____

d In the same school, how many would be expected to use neither service?

____

ISBN: 9780170370424

Two-way frequency tables

When there is data on two different characteristics for each member of a population, it is often convenient to display these in a two-way table.

Example: At another school of 714 students, each is asked about their use of Snapchat and Instagram. The results are shown in the table below.

	Use Instagram	Don't use Instagram
Use Snapchat	382	118
Don't use Snapchat	46	168

It is a good idea to add totals to your table:

	Use Instagram	Don't use Instagram	Totals
Use Snapchat	382	118	**500**
Don't use Snapchat	46	168	**214**
Totals	**428**	**286**	**714**

Check that both add to **714**.

a What is the probability that a student uses Snapchat?

$$P(\text{Snapchat}) = \frac{500}{714} = 0.7003$$

b What is the probability that a student uses neither Snapchat nor Instagram?

$$P(\text{neither}) = \frac{168}{714} = 0.2353$$

c If two students from this school are selected at random, what is the probability that both students use Snapchat?

One student AND the other use Snapchat so you MULTIPLY their probabilities.

$$P(\text{both students use Snapchat}) = \frac{500}{714} \times \frac{499}{713}$$
$$= 0.4901$$

One person has already been chosen.

d If an Instagram user is selected at random, what is the probability that he or she also uses Snapchat?

We are just concerned with the 428 Instagram users.

$$P(\text{Instagram user uses Snapchat}) = \frac{382}{428} = 0.8925$$

e Compare the answers to **a** and **d**. What do they tell you about users of Instagram and Snapchat at this school?

Students are more likely to use Snapchat (P = 0.8925) if they are also an Instagram user. The probability of using Snapchat in the group as a whole was only 0.7003.

 ISBN: 9780170370424

Use the data in the following tables to answer the questions.

1 Data was collected from the same school about how many Year 12 and 13 students went to the Formal.

	Year 12	Year 13	Totals
Went to the Formal	87	92	
Didn't go to the Formal	42	4	
Totals			

a What percentage of students went to the Formal?

b What is the probability that a randomly selected student was in Year 12 and went to the Formal?

c What is the probability that a Year 12 student went to the Formal?

d If two students were selected at random, what is the probability that both went to the Formal?

2 A survey was done of eight year old children in 2 towns. One town had fluoride added to its water and the other did not. The survey recorded whether or not there was evidence of tooth decay, for each child.

	Fluoride	No fluoride	Totals
No tooth decay	387	162	
Tooth decay	236	154	
Totals			

a What is the probability that a student had tooth decay?

b What is the probability that a child living in an area **without** fluoride had no tooth decay?

c What is the probability that a child living in an area **with** fluoride had no tooth decay?

d What is the probability that two randomly selected eight-year-olds from these areas both had tooth decay?

ISBN: 9780170370424

3 Theo did a survey of Year 11 students. He asked them whether or not they had ever been skiing.

	Been skiing	Had not been skiing	Totals
Boys	11	28	
Girls	16	29	
Totals			

a What is the probability that a randomly selected student had been skiing?

b What is the probability that a randomly selected student was a boy who had been skiing?

c What is the probability that a randomly selected girl had been skiing?

d What is the probability that two randomly selected students from his survey had both been skiing?

4 A recent survey of 1200 students (540 primary and 660 secondary) recorded whether they ate dinner at a table or not. The probabilities are shown in the table below.

	Primary	Secondary	Totals
Dinner at a table		0.30	
Dinner not at a table	0.18		**0.43**
Totals	**0.45**		

a How many of the students surveyed ate dinner at a table?

b How many students were from a primary school and did not eat dinner at a table?

c What is the probability that a primary student did not eat dinner at a table?

d How many students either were from a secondary school or ate dinner at a table?

ISBN: 9780170370424

5 The same survey of 1200 students (540 primary and 660 secondary) recorded whether they were born in New Zealand or not. The probabilities are shown in the table below.

	Primary	Secondary
Born in NZ	0.37	0.44
Not born in NZ	0.08	0.11

a How many of the students surveyed were not born in New Zealand?

b How many students were born in New Zealand and are from a secondary school?

c What is the probability that a secondary school student was born in New Zealand?

d How many students either were born in New Zealand or were at primary school?

6 The table below shows the numbers of passengers who survived and died on the *Titanic*, along with which class of ticket they had bought.

	First class	Second class	Third class	Totals
Survived	200	117	172	**489**
Died	119	152	527	**798**
Totals	**319**	**269**	**699**	**1287**

a What was the probability that a passenger died?

b What was the probability that a passenger was in first class and died?

c What was the probability that a first class passenger died?

d What was the probability that a third class passenger died?

e Explain what answers **a** — **d** tell you about the probability of dying on the Titanic.

ISBN: 9780170370424

Data handling

Collecting data

When collecting data, we nearly always have to take a **sample** from our **population**. We then use sample data to make **inferences** about the population.

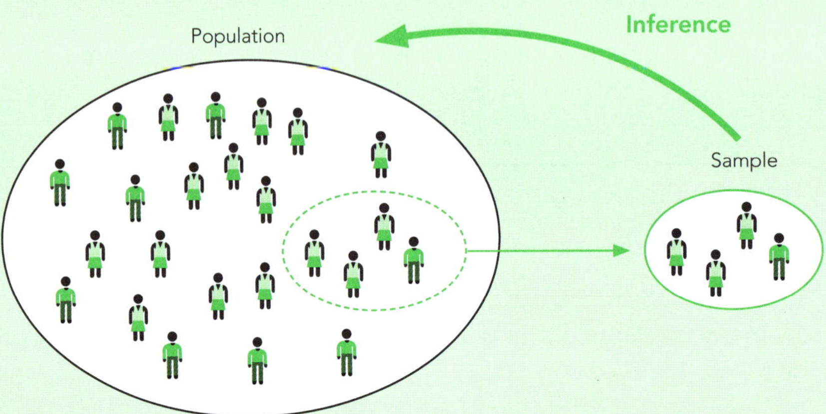

There are two important questions to ask when considering data from a sample:

1 Is the sample big enough?

A sample of 30 is generally considered enough for most purposes. However, the bigger the sample, the more confidence you can have of your findings.

2 Is the data biased?

Bias occurs when some members of the population are more likely than others to be selected for the sample, so the sample does not truly represent the population.
Examples:

- Ringing a radio station, filling in a form in a magazine or going to a website to give feedback: these are called 'self-selected' samples because the subject decides whether they will be sampled. Only those with an interest in the topic of the survey will be in the sample.
- The survey in the mall: this collects data only from those who go to malls, those who have time to answer, and there is an element of self-selection.
- Telephone surveys: are done on land lines, so mostly older people and people who don't often move house are surveyed. These also tend to be self-selected because many people choose not to respond.

Random sampling

In order to avoid bias, there needs to be a system of **random sampling** of the population. Every member of the population should have an equal chance of being selected.
Examples:

- Writing names on equal-sized pieces of paper and drawing them from a hat.
- Giving every member of the population a number and using random numbers to decide who will be in the sample.
- Using random numbers to decide who will be sampled from the electoral roll.
- Selecting every fifth sheep as they pass through a narrow gap.

ISBN: 9780170370424

State whether the following samples might be biased, and why.

1 In order to find the views of Year 10 students, the dean surveys one class of students.

2 In order to find the views of Paradise High School students, students walking past the senior common room are surveyed.

3 In order to find the views of Paradise High School students, every tenth student passing through the gate before 8.30 a.m. is surveyed.

4 In order to find the views of Paradise High School students, every tenth student on the school roll is surveyed.

5 In order to find the views of Paradise High School students about what food should be sold at the canteen, students queuing at the canteen at lunchtime are surveyed.

6 In order to find the views of Paradise High School students about what food should be sold at the canteen, response forms are left outside the student office, along with a drop box.

7 In order to find the views of Paradise High School students, all students with a family name starting with a particular letter of the alphabet are selected.

8 In order to find the views of Paradise High School students, 10 students are selected randomly from each year level.

ISBN: 9780170370424

Measures of centre

These give a measure of where the **middle** of a distribution lies.

Name	Calculation	Advantages	Disadvantages
Mean	$\frac{\text{the sum of all the data values}}{\text{the number of data values}}$	• Easy to calculate	• Distorted by very big or small values
Median	middle value	• Usually a good measure of centre	• Data must be put in order first
Mode	value that occurs most frequently	• Very easy to find	• Very unreliable as a measure of centre • Often there are two or none

Example: 45, 27, 39, 34, 97, 33, 46, 28, 22, 23, 50, 47, 48, 33, 37, 46, 19, 22

Put the values in order: 19, 22, 22, 23, 27, 28, 33, 33, 34, 37, 39, 45, 46, 46, 47, 48, 50, 97

$\text{Median} = \frac{34 + 37}{2}$

If there are >2 'modes', then there is no mode.

$\text{Mean} = \frac{696}{18} = 38.67$ $\text{Median} = \frac{34 + 37}{2} = 35.5$ Modal values = 22, 33 and 46 ∴ no mode

The best measure of centre is the median. Apart from the 97, all the data is fairly evenly spread between 19 and 50, and the midpoint of these is about 35, which is close to the median. The mean has been pulled up by just one value — the 97.

Calculate the mean, median and mode for the following sets of data, and state which, if any, you think is the best measure of centre and why.

1 1, 2, 5, 2, 4, 4, 3, 1, 1, 3

Mean = ____ Median = ____ Mode = ____

The best is the ____ because ____

2 18, 2, 19, 17, 14, 20, 16, 21, 20, 23

Mean = ____ Median = ____ Mode = ____

The best is the ____ because ____

 ISBN: 9780170370424

Measures of spread

These give a measure of how widely **spread** the data is.

Quartiles
The median is the middle value.
The upper quartile (UQ) is the middle value of the top half of the data.
The lower quartile (LQ) is the middle value of the bottom half of the data.

Interquartile Range (IQR) = UQ – LQ: This is a very good measure of spread because extreme values do not affect it.

Range = Maximum – Minimum: This is not a good measure of spread because it is affected by very big and very small values.

Example 1: 4, 9, 5, 8, 6, 0, 3, 8, 7, 1, 7, 5, 8, 2, 2, 6, 7, 6, 7, 0

Order the data:

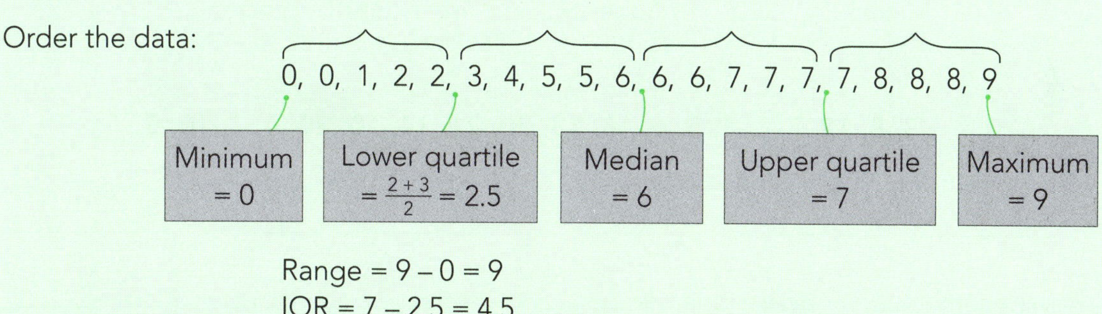

Range = 9 – 0 = 9
IQR = 7 – 2.5 = 4.5

Example 2: 4, 9, 5, 8, 6, 0, 3, 8, 7, 7, 5, 8, 2, 2, 3

Order the data:

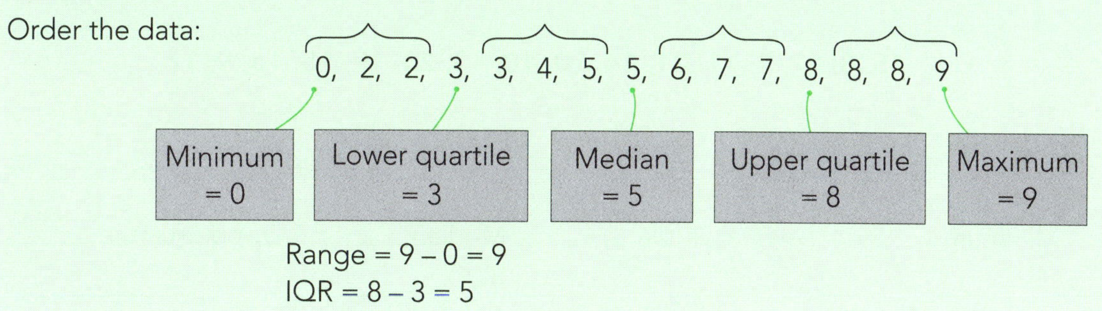

Range = 9 – 0 = 9
IQR = 8 – 3 = 5

Calculate the statistics for the following data sets.

1 5, 8, 10, 11, 14, 15, 15, 17, 18, 19, 20, 21, 22, 24, 25

____

Minimum ______ Lower quartile ______ Median ______ Upper quartile ______

Maximum ______ Range ______ Inter-quartile range ______

ISBN: 9780170370424

2 2, 5, 1, 15, 14, 10, 8, 9, 12, 14, 19, 6, 8, 10, 14, 7, 12

____

Minimum ______ Lower quartile ______ Median ______ Upper quartile ______

Maximum ______ Range ______ Inter-quartile range ______

3 1, 5, 0, 1, 1, 5, 2, 3, 3, 6, 8, 2, 2, 4, 5, 0, 0, 8

____

Minimum ______ Lower quartile ______ Median ______ Upper quartile ______

Maximum ______ Range ______ Inter-quartile range ______

4 3, 8, 12, 7, 4, 1, 5, 6, 18, 13, 20, 21, 14, 10, 20, 4, 7, 10, 6

____

Minimum ______ Lower quartile ______ Median ______ Upper quartile ______

Maximum ______ Range ______ Inter-quartile range ______

5 15, 8, 12, 7, 11, 21, 14, 10, 10, 13, 20, 19, 15, 16, 9, 12

____

Minimum ______ Lower quartile ______ Median ______ Upper quartile ______

Maximum ______ Range ______ Inter-quartile range ______

6 15, 18, 17, 13, 14, 20, 15, 17, 21, 15, 20, 20, 19, 17, 18, 14, 13

____

Minimum ______ Lower quartile ______ Median ______ Upper quartile ______

Maximum ______ Range ______ Inter-quartile range ______

ISBN: 9780170370424

Displaying data

1 Dot plots

A dot plot provides a visual representation of the data.

Example: 2, 3, 3, 5, 5, 5, 6, 7, 7, 8, 8, 8, 8, 9, 9, 9, 10, 10, 12, 13, 13, 14, 15, 18

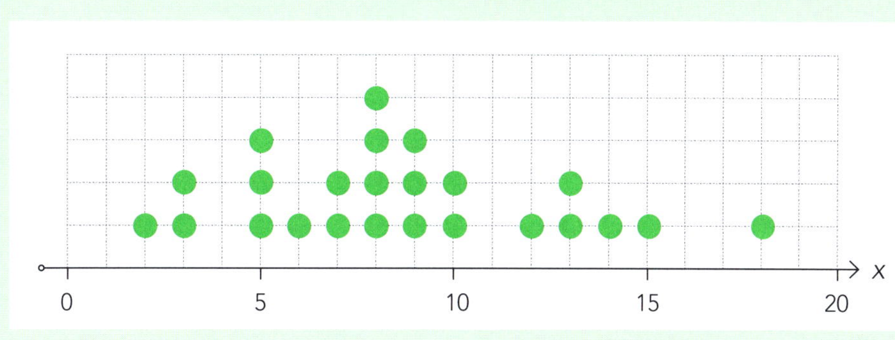

Create dot plots for these data sets.

1 2, 4, 4, 6, 7, 7, 8, 8, 9, 9, 9, 10, 10, 11, 11, 11, 12, 12, 13, 19

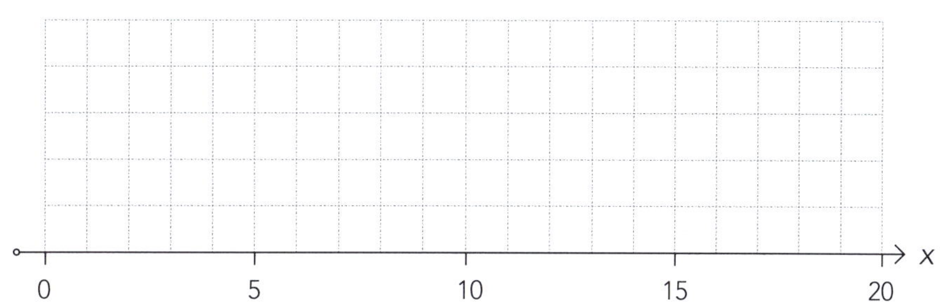

2 4, 5, 8, 15, 13, 11, 6, 12, 7, 8, 7, 13, 17, 12, 9, 8, 12, 13, 8, 13

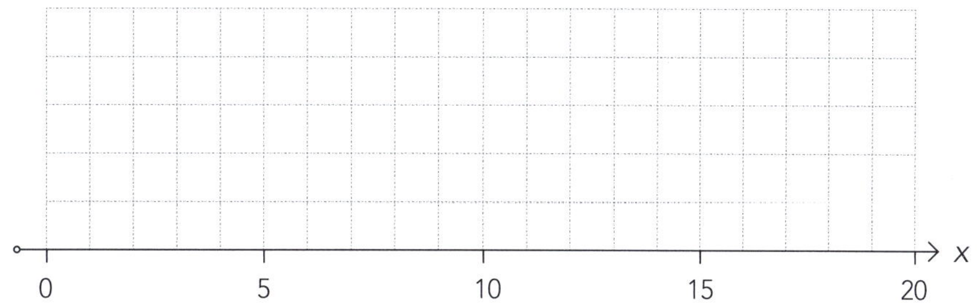

3 0, 5, 2, 8, 1, 6, 2, 2, 4, 0, 1, 9, 11, 3, 4, 2, 1, 1, 3, 3

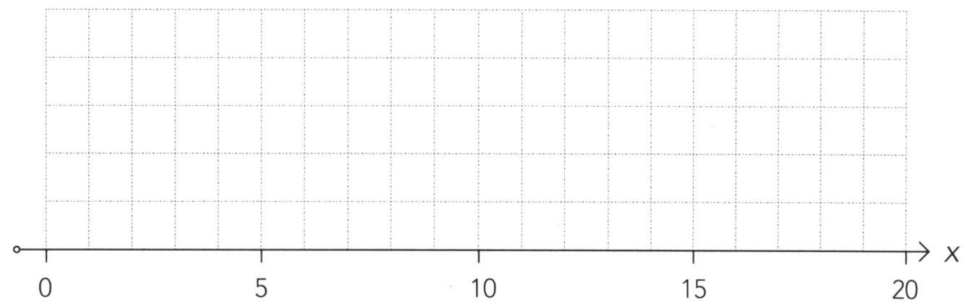

ISBN: 9780170370424

2 Box plots

- Box plots are excellent for displaying statistics and for comparing sets of data.
- However, they do not show the distribution of all the data points, for example a whisker may be very long because of just one extreme value, or there may be many extreme points.
- Each part of a box plot represents 25% of the data.

Before we can create a box plot, we require these statistics:

minimum lower quartile median upper quartile maximum.

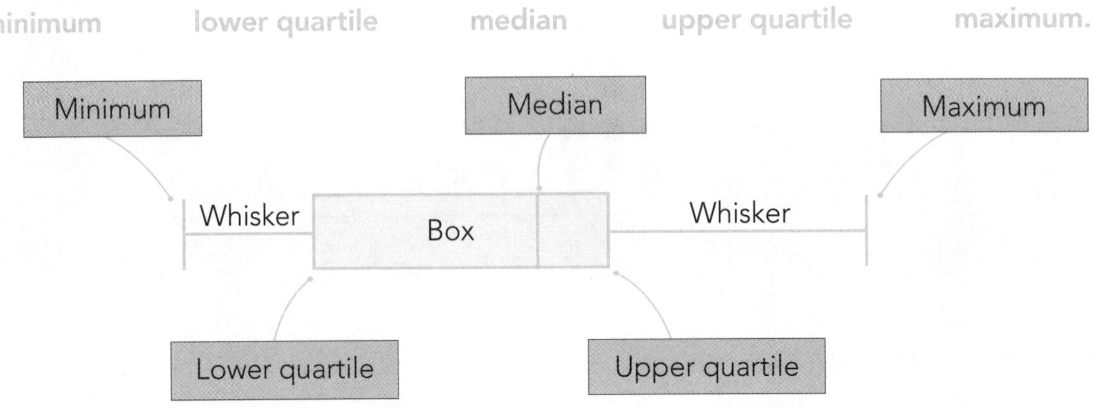

Example: 2, 3, 3, 5, 5, 5, 6, 7, 7, 8, 8, 8, 8, 9, 9, 9, 10, 10, 12, 13, 13, 14, 15, 18

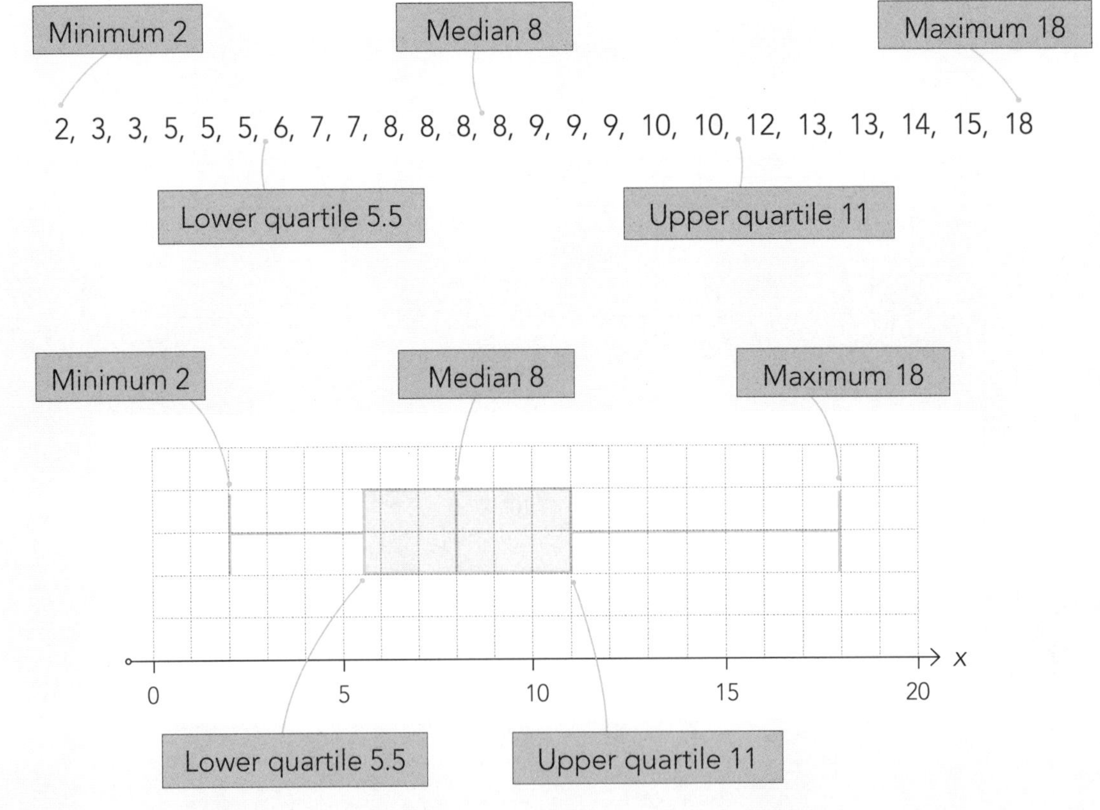

ISBN: 9780170370424

Draw box plots for the following sets of data.

1 2, 4, 4, 6, 7, 7, 8, 8, 9, 9, 9, 10, 10, 11, 11, 11, 12, 12, 13, 19

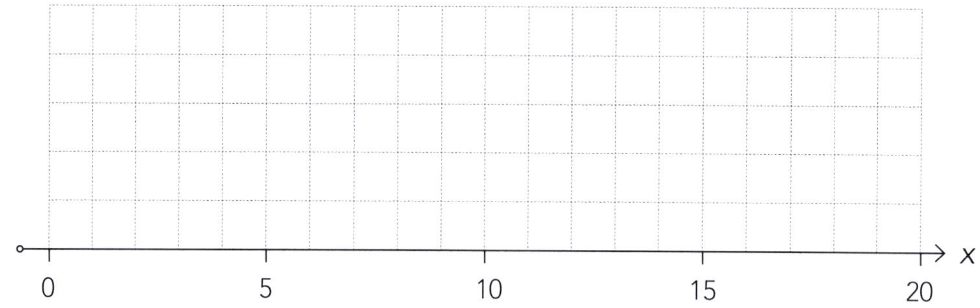

2 4, 5, 8, 15, 13, 11, 6, 12, 7, 8, 7, 13, 17, 12, 9, 8, 12, 13, 8, 13

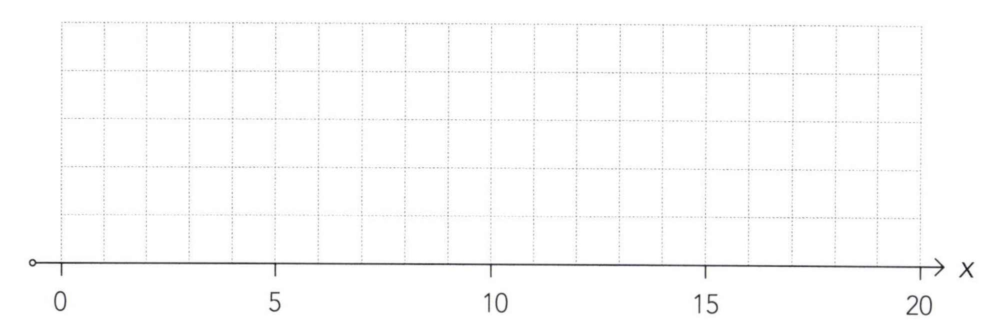

3 0, 5, 2, 8, 1, 6, 2, 2, 4, 0, 1, 9, 11, 3, 4, 2, 1, 1, 3, 3

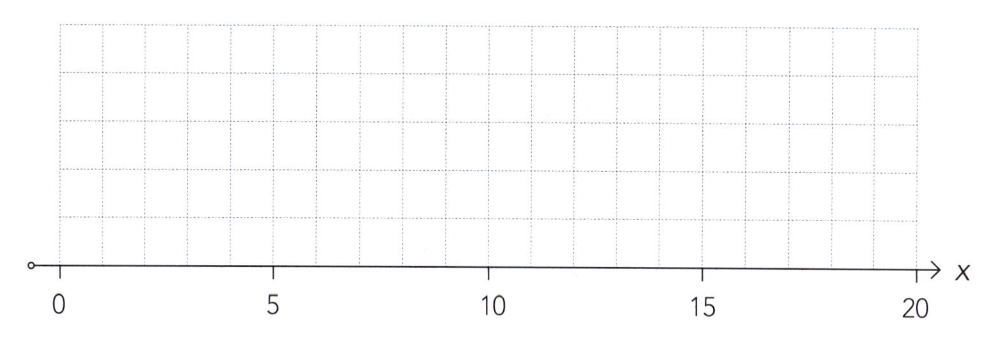

4 6, 5, 7, 8, 1, 6, 2, 12, 4, 8, 4, 9, 11, 13, 4, 2, 7, 10, 3, 3

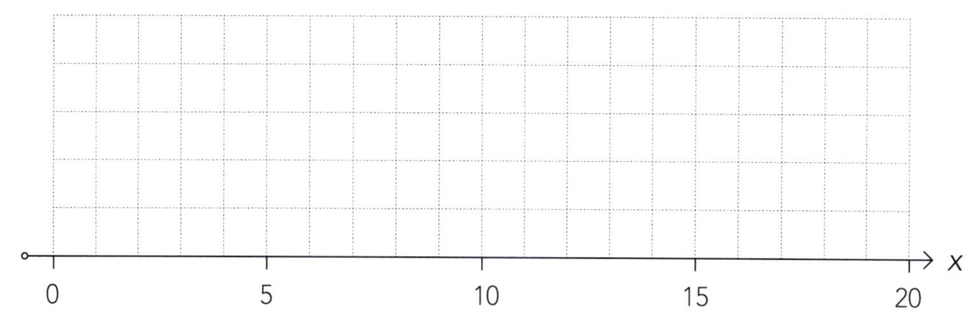

ISBN: 9780170370424

Combining dot plots, box plots and means

The first three examples in the previous exercises have used the same data. The answers have been put on the same graph so you can see the relationship between them.

1

Mean = 8.95

- The mean (8.95) and the median (9) are very close because the data is distributed fairly symmetrically around the mean.
- Notice that the length of the whisker on the right is much longer because of just one value.
- The gaps within the box are much smaller than the lengths of the whiskers because the data is clustered in the centre.

2

Mean = 10.05

- The data is fairly symmetrically spread around the median, so the median (10) and the mean (10.05) are close.
- The dot plot shows that this data is 'bimodal' — it has two modes. This is **not** shown by the box and whisker plot.
- Because of the gap in the middle of the data, the minimum, lower quartile, median, upper quartile and maximum are fairly evenly spread.

3

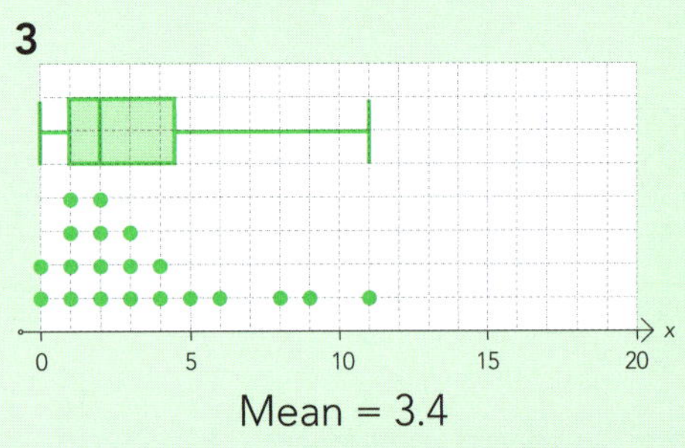

Mean = 3.4

- Most of the data is clustered on the left, which means the gaps between the minimum, lower quartile and median are small.
- The mean (3.4) is bigger than the median (2) because the 'tail' of bigger numbers on the right increases the mean, but not the median.
- The data is skewed to the **right**.

Match the dot plots with the box plots and delete words to make a true statement about the mean.

1

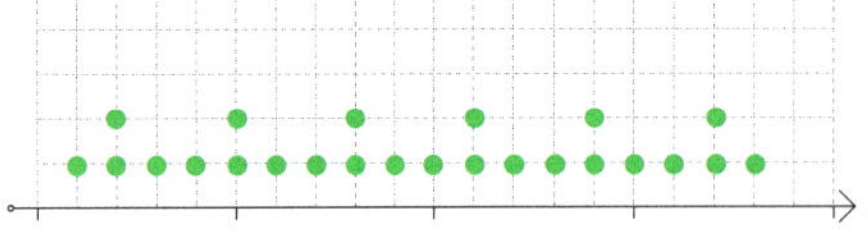

Mean is above/below/about the same as the median.

a

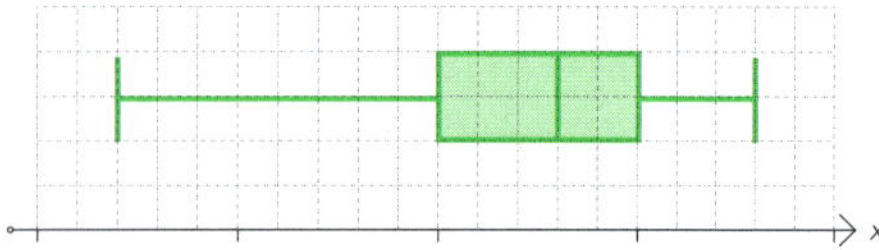

2

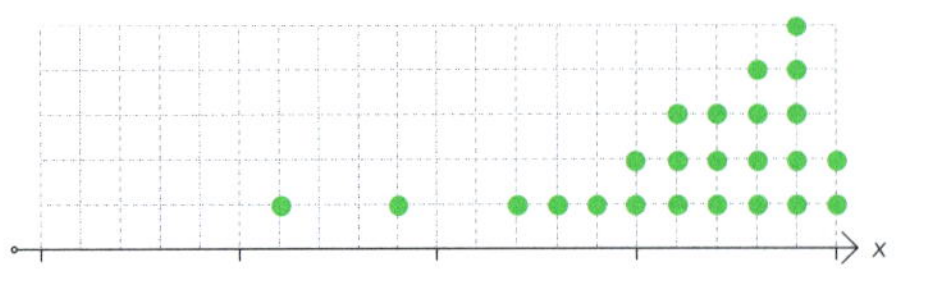

Mean is above/below/about the same as the median.

b

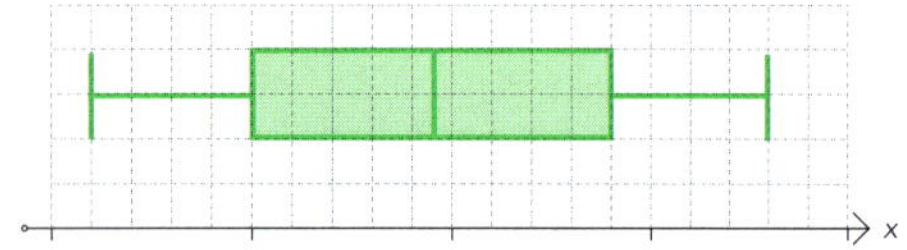

3

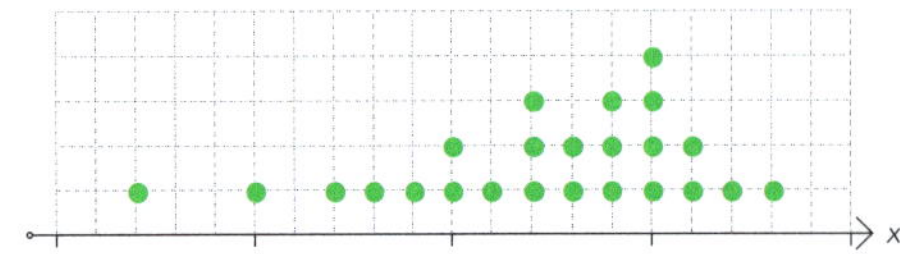

Mean is above/below/about the same as the median.

c

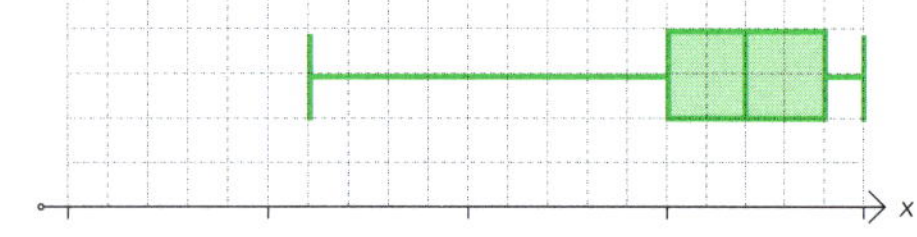

ISBN: 9780170370424

Drawing conclusions from pairs of box plots

- We can conclude that there is a difference between samples by comparing their box plots.
- In order to conclude that there might be a difference between populations by comparing their box plots, we need to consider the positions of the medians and middle 50% of both samples.

In all three cases below, the box plots show the lengths of individuals in samples from populations A and B.

Case 1: The boxes of each sample overlap with each other, and both medians lie within the box of the other sample.

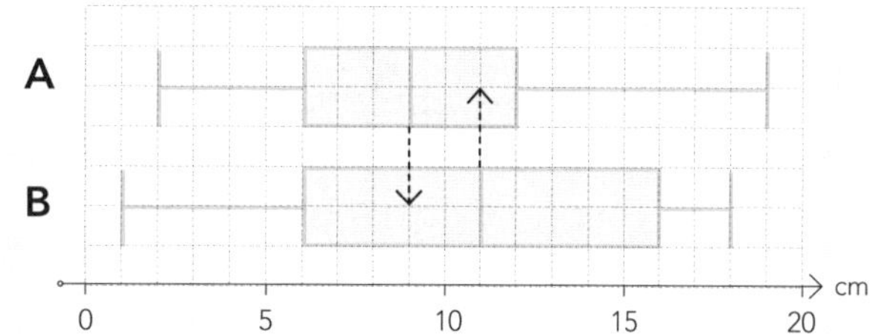

Observations: For the sample:

Comment on centre.

The lengths in sample B tend to be greater than those in sample A because the box in B extends further to the right than that of A. This is supported by the median of B (11) being bigger than the median of A (9).

Comment on spread.

Although the range is the same for both samples (17), the IQR for B (10) is larger than that for A (6), so the lengths of the middle 50% in B are more spread than those in A.

For the population:
We cannot conclude that individuals in population B tend to be longer than individuals in population A, because both of the medians lie within the middle 50% of the other sample, and the boxes overlap.

Case 2: The boxes of each sample overlap with each other, and one or both medians lie outside the box of the other sample.

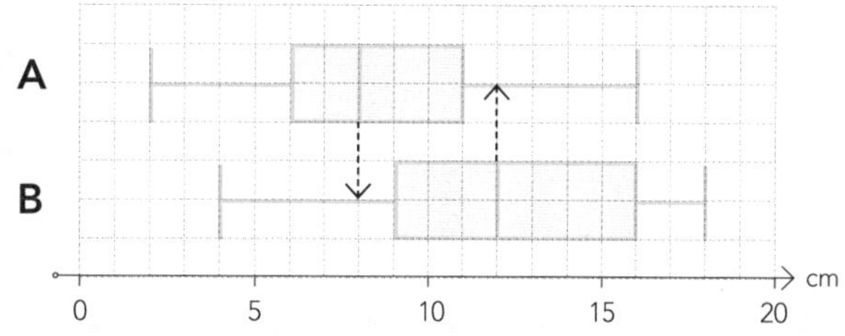

Observations: For the sample:

Comment on centre.

Individuals in sample B tend to be longer than those in sample A because the box in B is further to the right than that of A. This is supported by the median of B (12) being bigger than the median of A (8).

Comment on spread.

Although the range is the same for both samples (14), the IQR for B (7) is larger than that for A (5), so the lengths of the middle 50% in B are a little more spread than those in A.

For the population:
We can conclude that individuals in population B are likely, on average, to be longer than those in population A because, although the middle 50% of each sample overlap, both medians lie outside the boxes of the other group.

ISBN: 9780170370424

Case 3: The boxes of each sample do not overlap with each other, so both medians lie outside the box of the other sample.

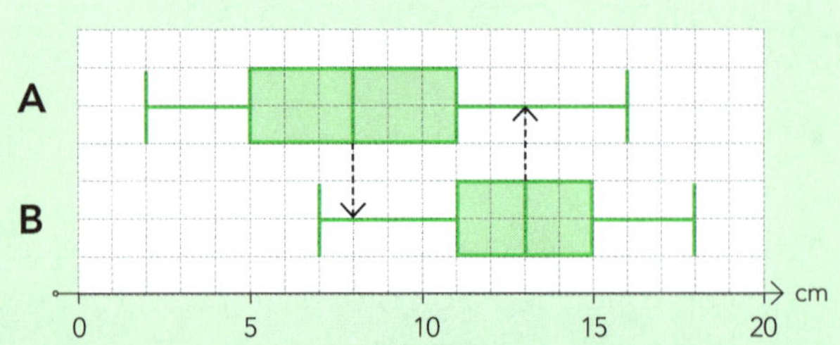

Comment on centre.

Comment on spread.

Observations: For the sample:
Individuals in sample B tend to be longer on average than those in sample A because the box for B is further to the right than that of A. This is supported by the median of B (13) being bigger than the median of A (8). The IQR for B (4) is larger than that for A (6), so the heights in sample B are more spread than those of A.

For the population:
We can conclude that individuals in population B will tend to be longer on average than those in population A because the boxes do not overlap and both medians lie outside the other box.

Write conclusions about the heights of individuals in the populations for each of the pairs of sample box plots.

1

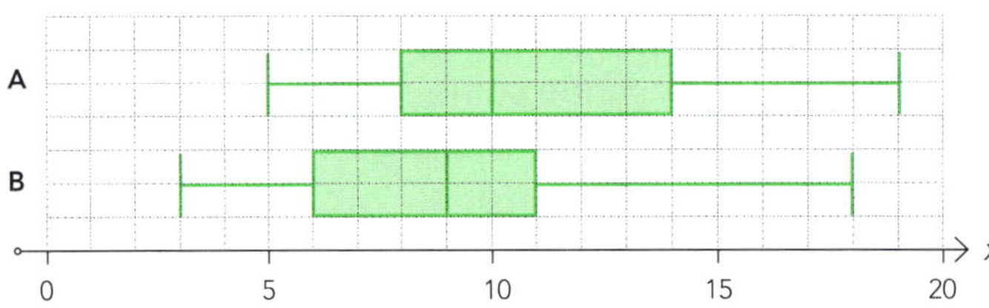

2

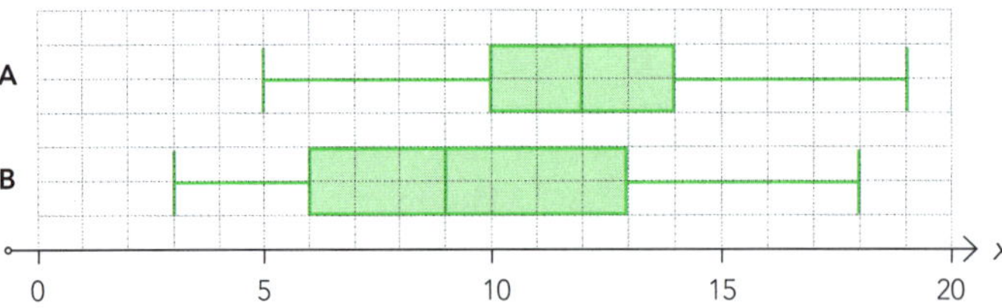

3

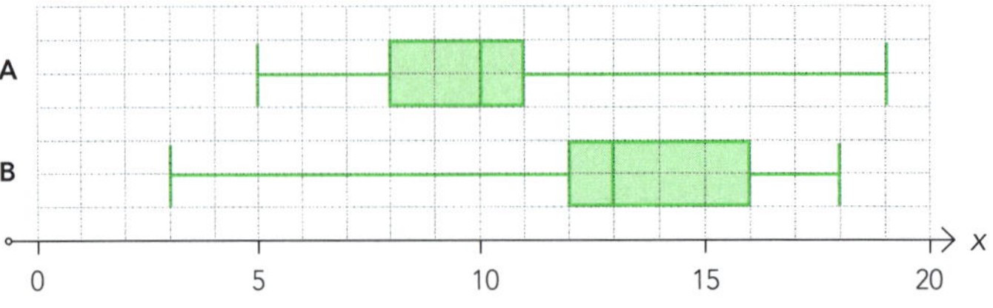

ISBN: 9780170370424

Comparing groups using dot and box plots

Example: One class of Year 11 students got results back from recent tests. Both were marked out of 15. The same tests were sat by all Year 11 classes at the school.

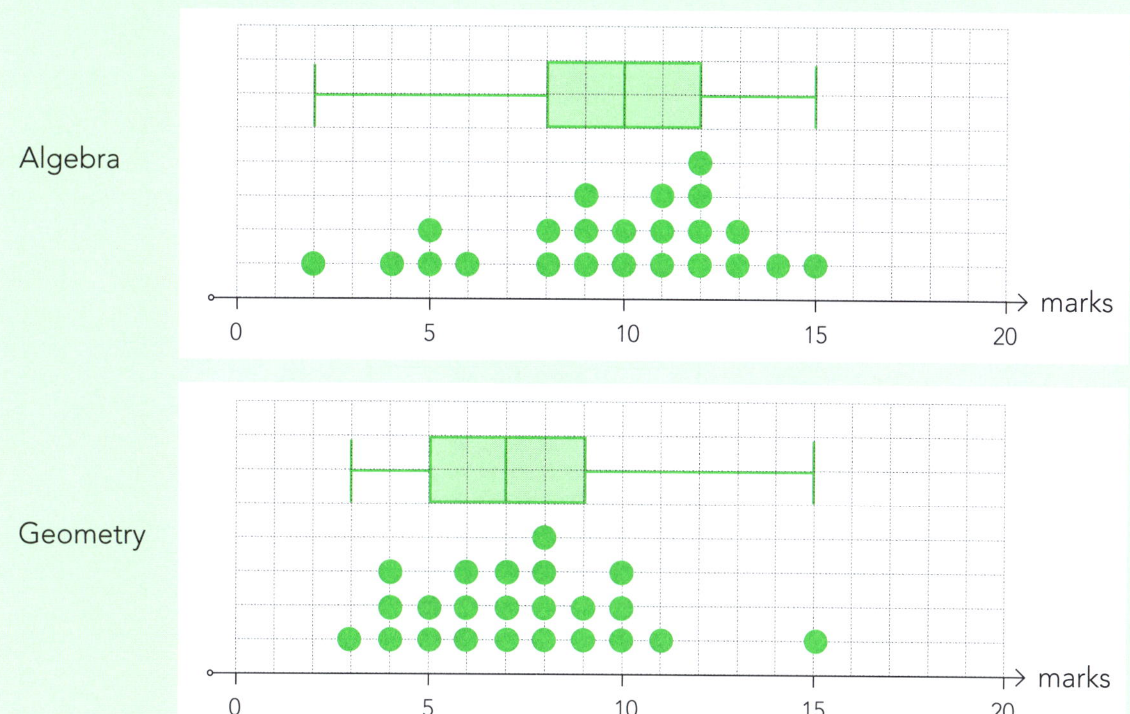

	Algebra	Geometry
Minimum	2	3
Lower quartile	8	5
Median	10	7
Upper quartile	12	9
Maximum	15	15
Mean	9.61	7.39

a What is the difference between the medians for the two tests?

Difference = 10 – 7 = 3

b Did this class of Year 11 students do better in the Algebra or Geometry test?

The dot plot suggests that this class of Year 11 students did better on average in the Algebra test than in the Geometry test because the Algebra dots are further to the right. This is supported by both the medians (Algebra 10, Geometry 7) and the means (Algebra 9.61, Geometry 7.39).

c Would you expect to come to the same conclusion for the entire group of Year 11 students?

Because both the medians lie outside the boxes of the other group, but the boxes overlap, there is evidence to suggest that all the Year 11 students who sat the tests are likely to do better on average in Algebra than in Geometry. However, this class might not be typical of the entire group. Their teacher might have spent more time on Algebra than other teachers, or it may be some time since they studied Geometry.

ISBN: 9780170370424

Answer the following questions.

1 The same class of Year 11 students sat similar tests the following term. Both were marked out of 15. Once again, the same tests were sat by all Year 11 classes at the school.

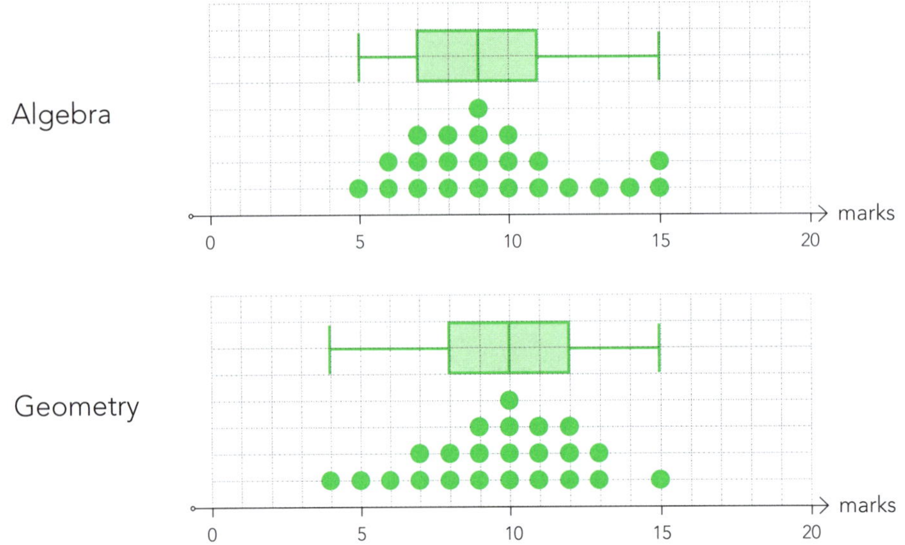

	Algebra	Geometry
Minimum		4
Lower quartile		8
Median		10
Upper quartile		12
Maximum		15
Mean	9.52	9.65

a Complete the table.

b What is the difference between the means for Algebra and Geometry?

____

c Do you consider that the students did better in one of the tests?

____

____

d Would you expect to come to the same conclusion for the entire group of Year 11 students?

____

____

e If you wanted to compare the results for these tests for the whole of Year 11, suggest a better way of selecting a sample.

____

____

 ISBN: 9780170370424

2 Thirty Year 11 boys and girls were chosen as they walked past the library, and asked to type the same text message into their cellphones. They were timed to see how long they took to type the message.

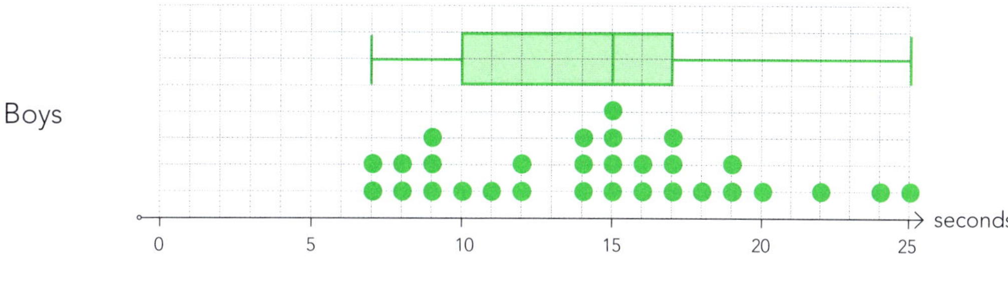

Girls

0 5 10 15 20 25 seconds

	Boys	Girls
Minimum	7	8
Lower quartile	10	12
Median	15	14
Upper quartile	17	16
Maximum	25	24
Mean	14.5	14.2

a Draw the box plot for the girls' times.

b Who texted the fastest, boys or girls? Justify your answer.

c Would you expect to come to the same conclusion for the entire group of Year 11 students?

d Calculate the inter-quartile ranges for both groups. What do these tell you about the distributions?

e How could you improve this investigation?

3 When there is doubt about who wrote a passage of prose, analysing the number of words in sentences can be used to distinguish authors. The lengths of 30 randomly selected sentences written by Katherine Mansfield in her short stories, and from a book by Jane Austen are shown on the graphs below.

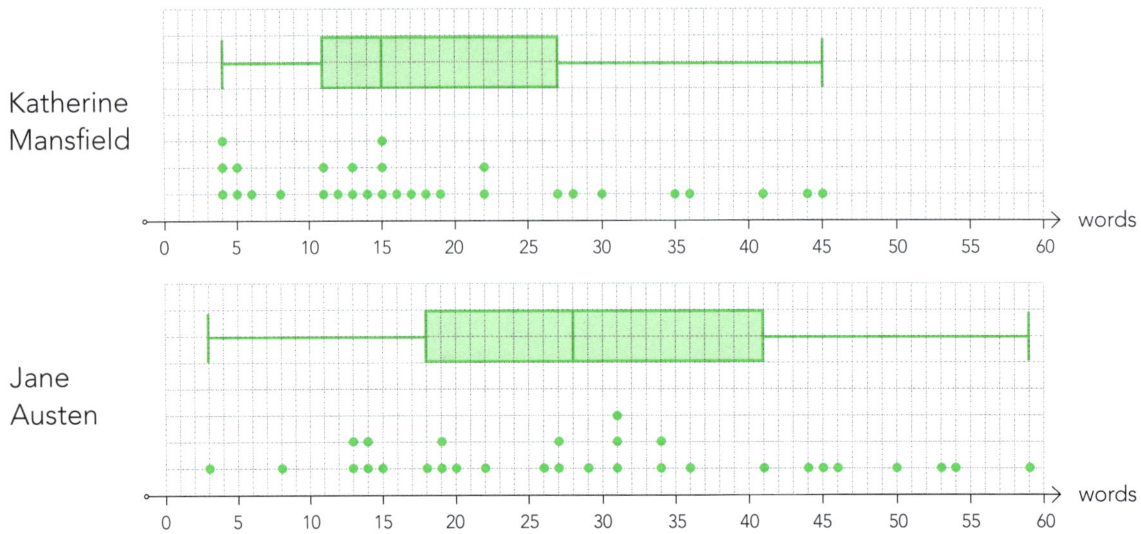

	Katherine Mansfield	Jane Austen
Minimum		3
Lower quartile		18
Median		28
Upper quartile		41
Maximum		59
Mean	18.5	29.2

a Complete the table.

b What is the difference between the medians?

____

c Who wrote longer sentences?

____

d Would you expect to come to the same conclusion if an analysis was done on all their work?

____

e Why is the mean for the length of Katherine Mansfield's sentences different from the median?

____

____

f How could you improve this investigation?

____

____

 ISBN: 9780170370424

4 It is also possible to distinguish prose written by different authors by analysing the lengths of words used. The graphs below show the lengths of 30 randomly selected words used in a book by Jane Austen and a book by Charles Dickens.

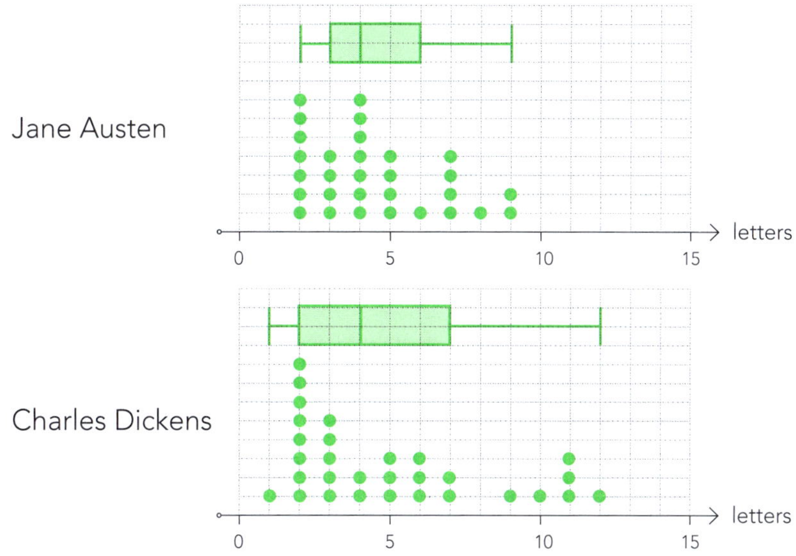

	Jane Austen	Charles Dickens
Minimum	2	1
Lower quartile	3	2
Median	4	4
Upper quartile	6	7
Maximum	9	12
Mean	4.67	5.06

a What is the difference between the mean word lengths?

____

b Is there a difference between the lengths of words used by these two authors?

____

____

c Would you expect to come to the same conclusion if an analysis was done on all their work?

____

____

d Whose word lengths were most consistent?

____

e How could you improve this investigation?

____

____

ISBN: 9780170370424

Bivariate data

- This is where you have **two** pieces of **numeric** (number) data about every individual in a sample or population.
- You analyse **both** at once to see if there is a relationship between them.
- These are plotted on a **scatter** graph.

Line of best fit

- It is useful to draw a line of best fit before discussing a bivariate relationship.
- This line should – have about the same number of points above and below it
  – pass through the middle of the points at both ends.

Examples:

Good line of best fit:

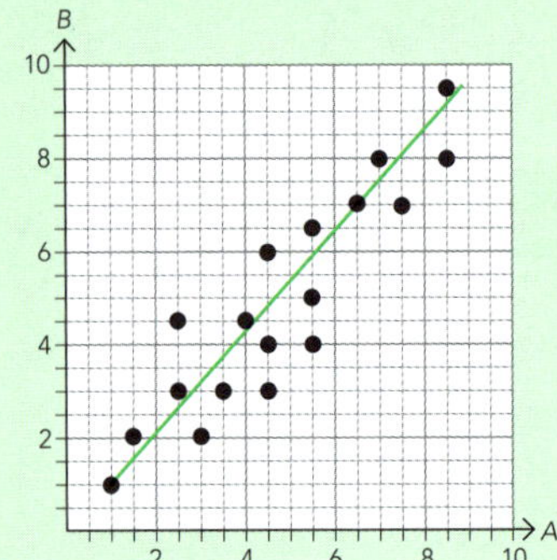

Poor lines of best fit:

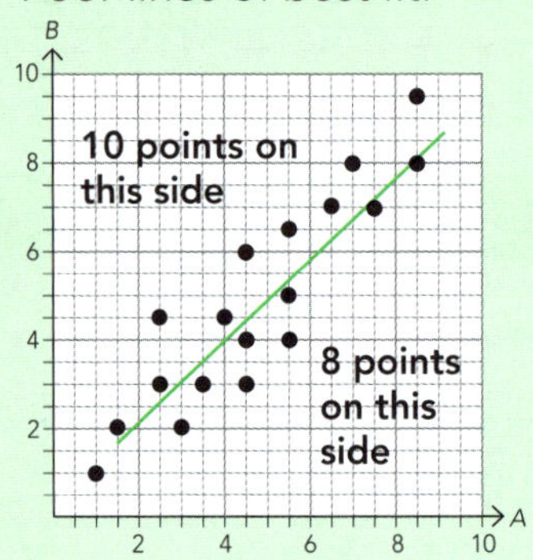

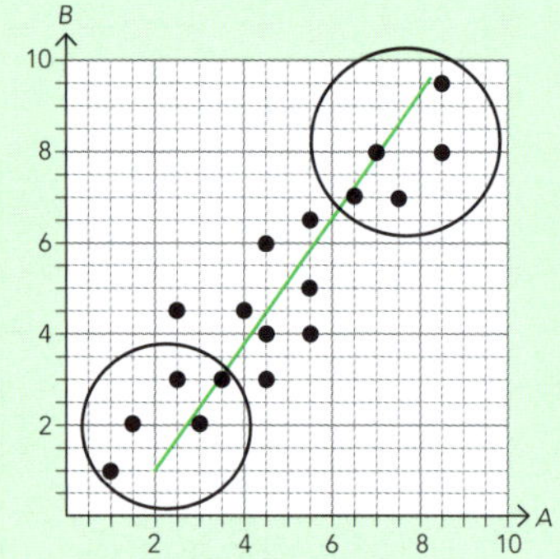

8 on both sides but not through the middle at both ends.

Features of scatter graphs, which need to be considered

1 Is the relationship positive or negative?

Positive:

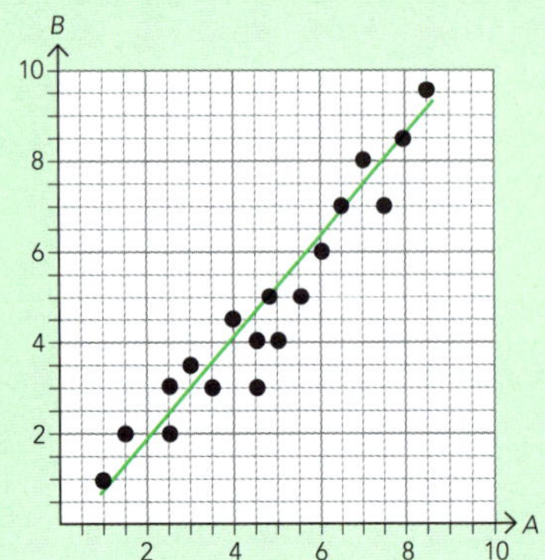

As A gets bigger, B gets **bigger**.

Negative:

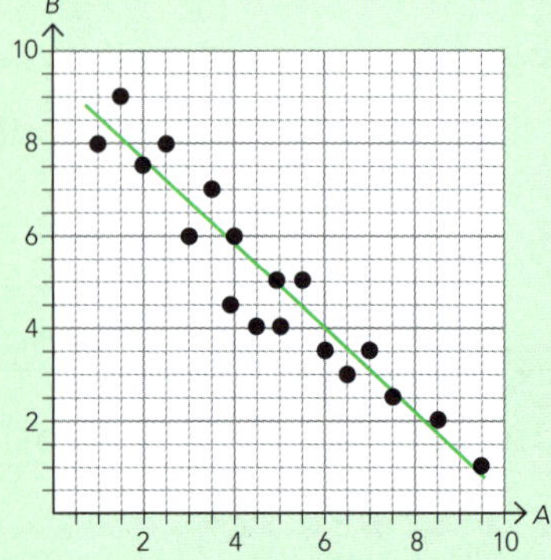

As A gets bigger, B gets **smaller**.

ISBN: 9780170370424

2 Is the relationship strong or weak?

Strong:

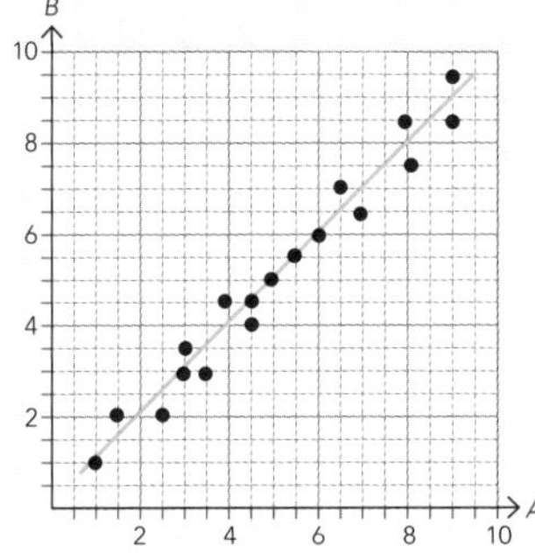

This means:

- most points are close to the line of best fit
- you can have confidence that the relationship is strong.

Moderate:

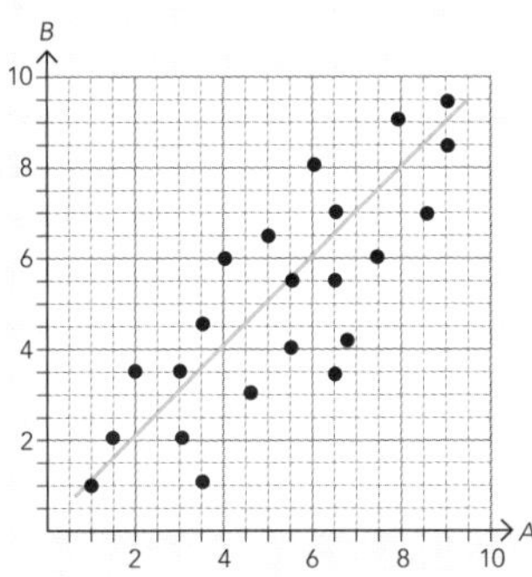

This means:

- most points are reasonably close to the line of best fit
- the relationship is moderately strong.

Weak:

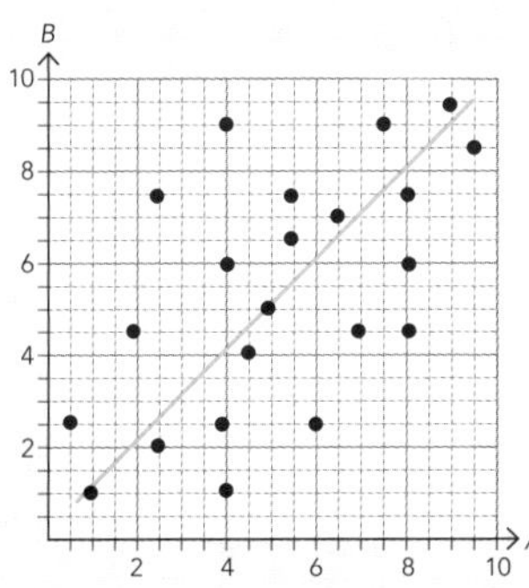

This means:

- many points are some distance from the line of best fit
- the relationship is weak.

Reading a graph value

This is done in the same way as for any other graph: (x-coordinate, y-coordinate)
or: (horizontal, vertical).

Examples:

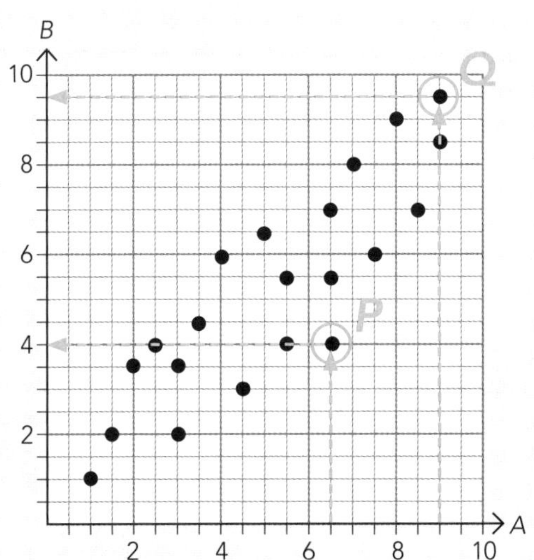

Point P: (6.5, 4)

Point Q: (9, 9.5)

ISBN: 9780170370424

Unusual points

- These are points that lie some distance from most of the other points.
- Sometimes these are valid points.
- Otherwise they may be a result of measurement error, recording error, or reversed coordinates.

Examples:

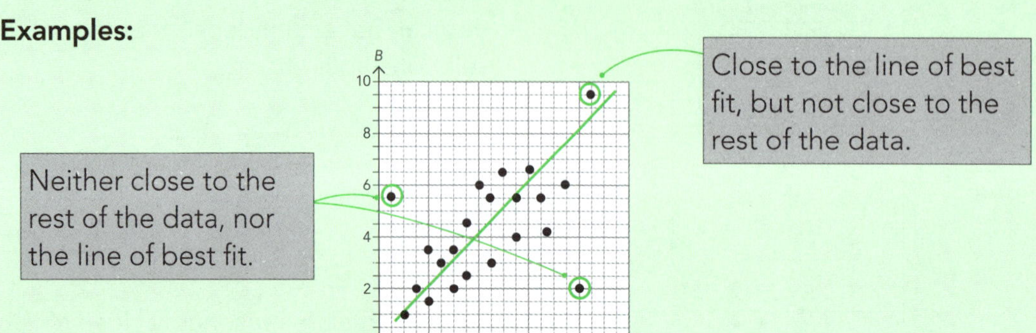

Putting it together

Example: Tara sells cups of soup at the local market during the winter. She records the air temperature and how many cups she sells each day. Her data is shown below.

a Draw a line of best fit.

b How many cups of soup did she sell on the day it was 16°C?
24 cups of soup.

c On how many days did she sell more than 40 cups?
On 10 days.

d Are there any unusual points?
(24, 16): Unusually warm at 24°, and she sold only 16 cups of soup.

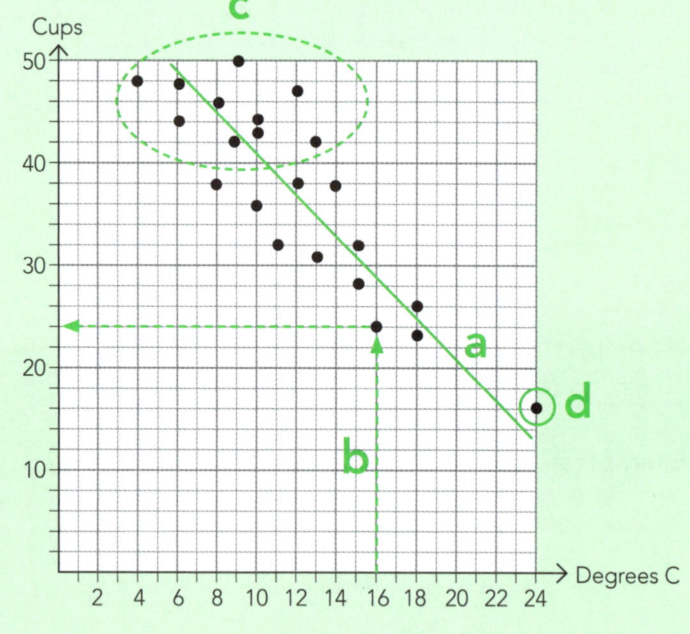

e Why is a scatter graph appropriate for this data?
Because for every day there are two pieces of numeric data: the temperature and the number of cups of soup sold.

f Describe the relationship.
There is a negative and reasonably strong relationship between the temperature and the number of cups of soup sold. The relationship is reasonably strong because most points are close to the line of best fit. It is negative because the higher the temperature, the fewer the number of cups of soup sold.

ISBN: 9780170370424

Answer the following questions.

1 The graph below shows the ages (to the nearest year) of the man and the woman for 26 couples.

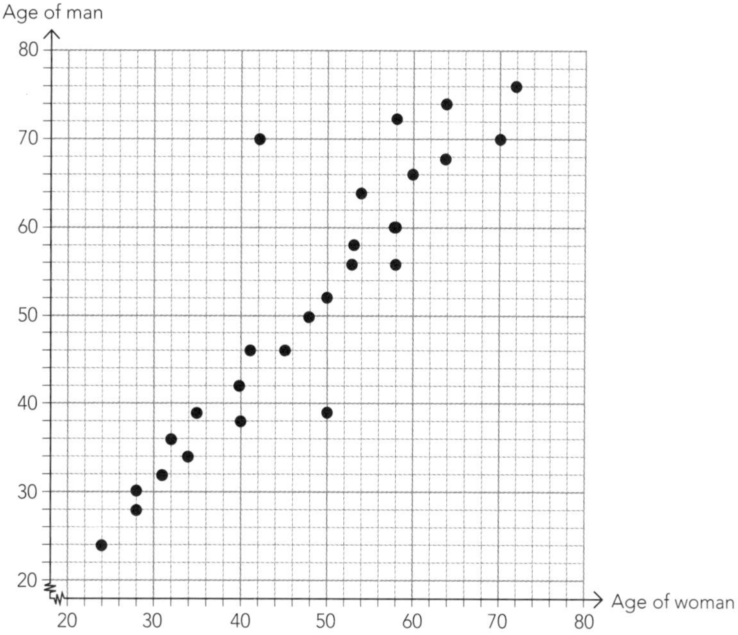

a Draw and label the line of best fit.

b The woman in one couple was 72. How old was her husband?

____

c In how many couples was the man younger than 40?

____

d Do you consider there are any unusual points? If so, give the coordinates and explain why.

____

____

e Draw and label a line to show where all the points would be if the ages of men and women in couples were the same.

f Give the ages of three couples in which the woman is older than the man.

____

____

g Describe the relationship between the ages of men and women in couples.

____

____

____

h Explain why a scatter graph is appropriate for this data.

____

____

ISBN: 9780170370424

2 26 children of different ages were given a puzzle. The time each took to complete the puzzle was recorded. The results are shown below.

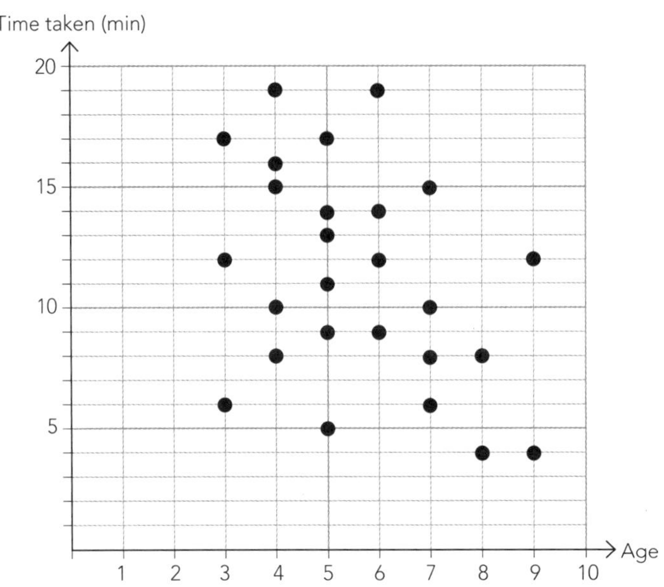

a Draw a line of best fit.

b There were two eight-year-olds. How long did they take to complete the puzzle?

____

c How many children took 6 minutes to complete the puzzle?

____

d Do you consider there are any unusual points? If so, give the coordinates and explain why.

____

____

____

____

e Describe the relationship between age and time taken to complete the puzzle.

____

____

____

____

____

f How could you improve this investigation so that you could be more confident of your answer?

____

____

____

ISBN: 9780170370424

3 The graph shows the relationship between the percentage of packaging paper and packaging glass that is recycled by a range of countries.

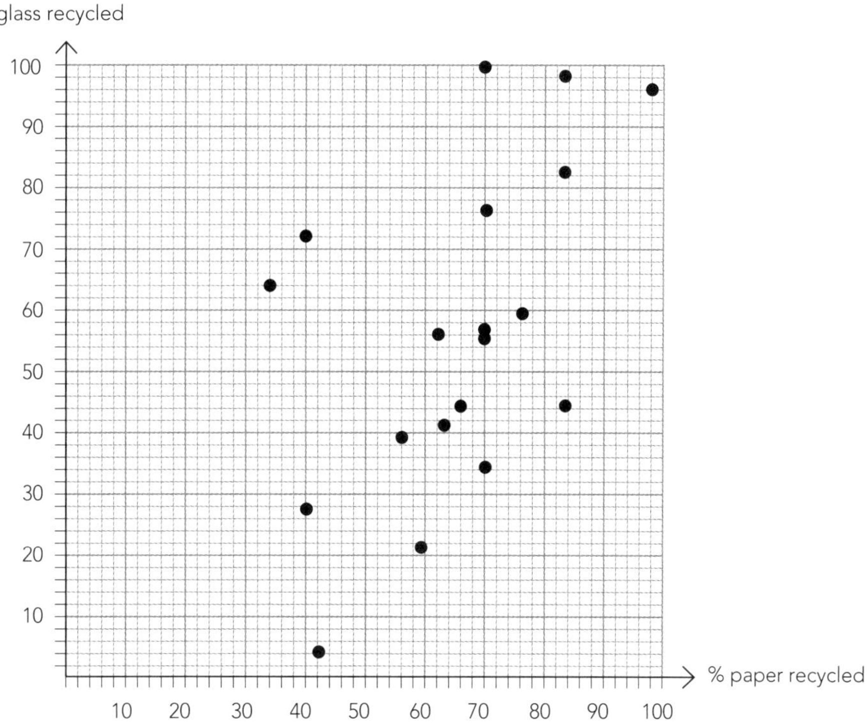

a Draw and label a line of best fit.

b How many countries recycle less than half their paper packaging?

____

c One country recycles 34% of its packaging paper. How much of its packaging glass does it recycle?

____

d Do you consider there are any unusual points? If so, give the coordinates and explain why.

____

____

e Draw and label a line for countries that recycle the same percentage of paper packaging as glass packaging.

f Do countries tend to recycle more paper or more glass packaging?

____

____

g Describe the relationship between the recycling of paper and glass packaging.

____

____

____

ISBN: 9780170370424

4 The graph below shows the relationship between average body mass and maximum lifespan for different breeds of dogs.

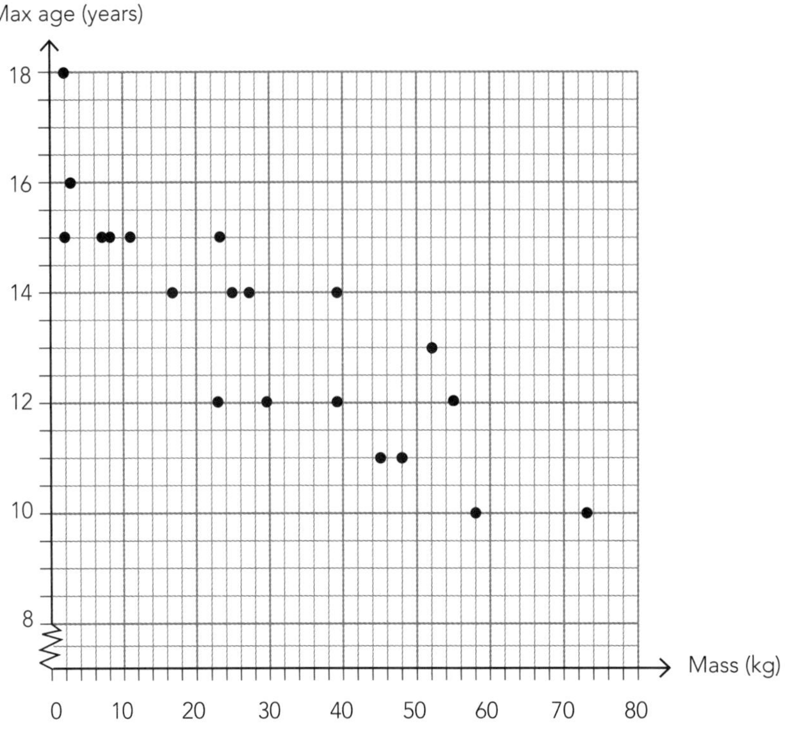

a Draw a line of best fit.

b What is the maximum lifespan of a breed that has an average mass of 52 kg?

____

c Give the average mass(es) of any species of dog that has/have a maximum lifespan of 11 years.

____

d Explain why a scatter graph is appropriate for this data.

____

____

____

e Do you consider there are any unusual points? If so, give the coordinates and explain why.

____

____

f Describe the relationship between the average mass and maximum lifespan for breeds of dogs.

____

____

____

ISBN: 9780170370424

5 The graph below shows the body length of some species of birds (mm), and the incubation period for their eggs (the number of days between laying the egg and hatching).

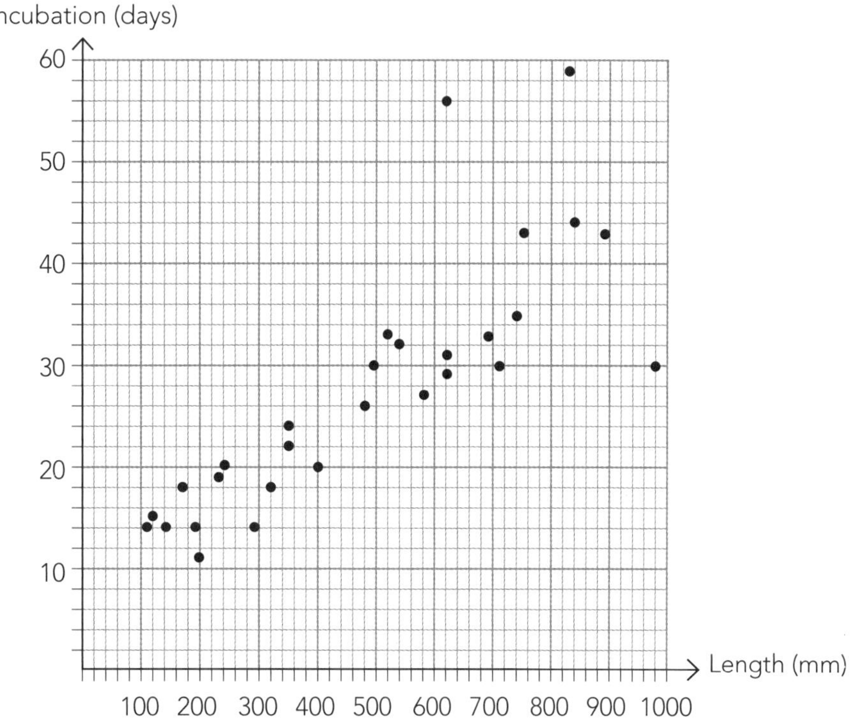

a Draw a line of best fit.

b The mallard duck is 580 mm long. What is its incubation period?

____

c The kingfisher has an incubation period of 19 days. How long is it?

____

d Do you consider there are any unusual points? If so, give the coordinates and explain why.

____

____

____

e Describe the relationship between the sizes of birds and their incubation periods.

____

____

____

f What could you do to improve this investigation?

____

____

____

Time series

- This is a series of values collected at **regular intervals of time**.
- The intervals could be years, months, weeks, days, hours, minutes, seconds or even milliseconds.
- These are plotted on a **line graph**.

Features of time series

1 The trends

These describe any **overall** change. There may be an overall increasing or decreasing trend. There may be a change in the amount of variation around the trend line.

Examples:

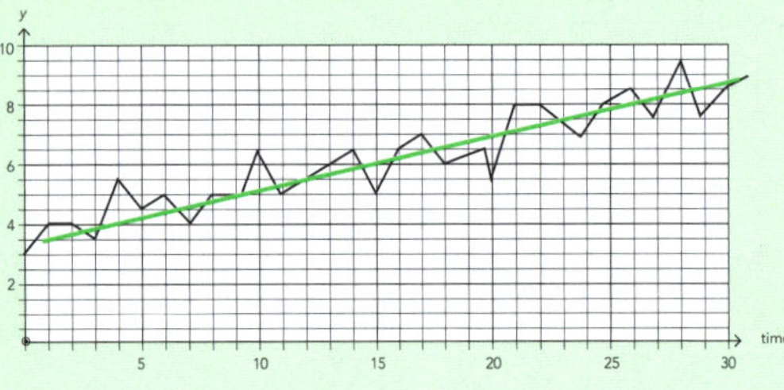

Increasing trend.
Variation around the trend line does not change.

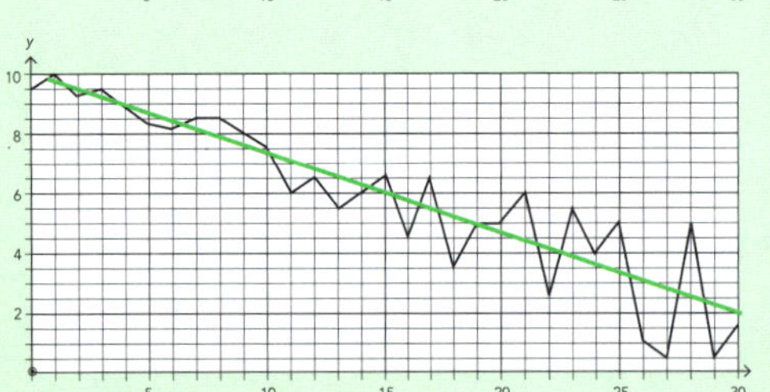

Decreasing trend.
Variation around the trend line increases.

2 The cycle

This is where a **repeating pattern** occurs at **regular** intervals along the time series.

This may be due to: annual changes, e.g. seasonal change in weather
weekly changes, e.g. effects of weekends on traffic flows
daily changes, e.g. effects of light on photosynthesis in plants etc.

Example:

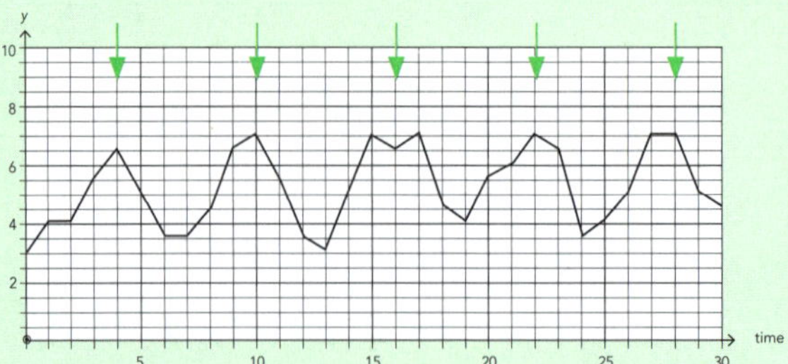

Notice that the cycle is not perfect. The peaks are about 6 units apart, and of fairly consistent heights. The low points are about halfway between the peaks, but their heights are more variable.

3 Error, noise, random — different words for the same meaning

These are small changes in the series due to **unpredictable** factors. This stops the series and cycles being perfectly smooth and regular.

ISBN: 9780170370424

Single time series

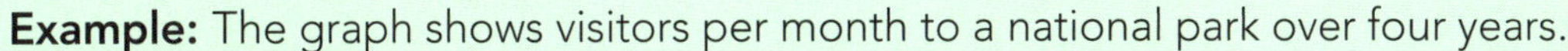

Example: The graph shows visitors per month to a national park over four years.

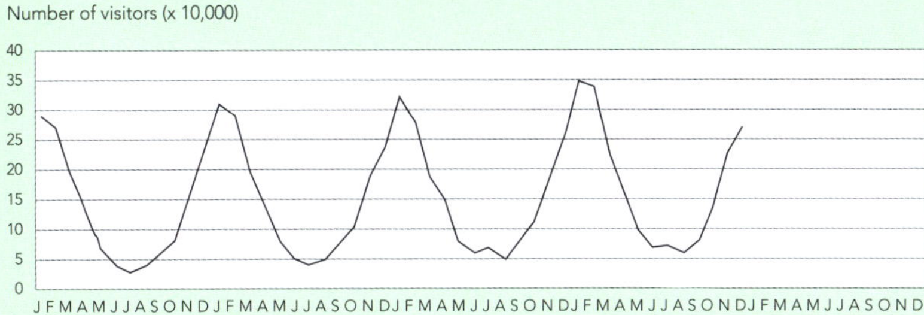

a Draw a trend line on the graph.

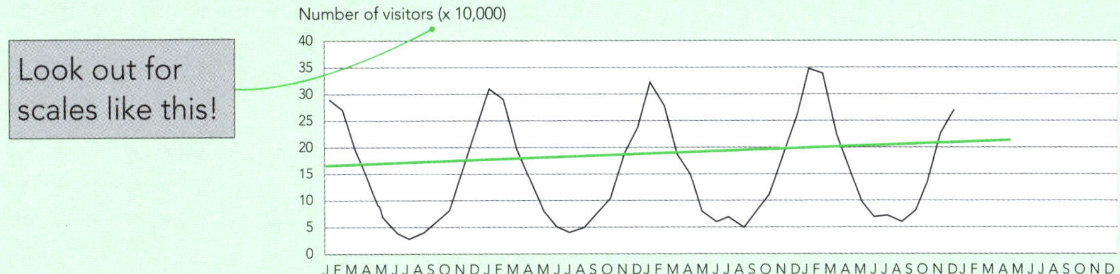

b Describe the trends.

Overall the number of visitors to this national park increases steadily by about 10 000–20 000 people per month over the four years shown. This is shown clearly by the increasing peaks in January, and by the increasing troughs in June or July. The variation around the trend line does not change.

c Describe the repeating pattern.

The highest number of people visit the park in January (290 000–350 000), and the second highest in February. The numbers decrease after February until they reach the lowest in June, July or August (40 000–70 000), and from then they increase rapidly until they return to a peak in January again.

d On the graph, sketch a prediction for the following 12 months.

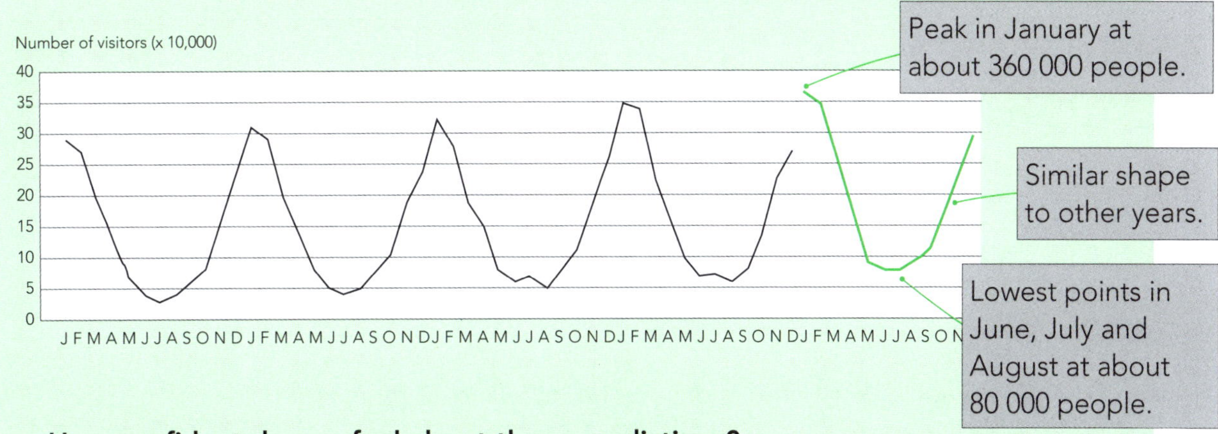

e How confident do you feel about these predictions?

I feel reasonably confident because the number of visitors has increased quite steadily by about the same number each year. However, there is only four years' data, which is not a lot. Also, numbers of visitors is likely to be influenced by unseasonal weather, how well the economy is doing, exchange rates, etc.

ISBN: 9780170370424

Answer the following questions.

1 The graph shows the number of passengers travelling on a ferry each month during a four-year period.

a Draw a trend line on the graph

b Estimate the number of passengers who travelled on the ferry during the first April of the given data.

____

c Estimate the number of passengers who travelled on the ferry during the last April of the given data.

____

d Describe any trends.

____

____

____

____

e Describe the repeating pattern.

____

____

____

____

f On the graph, sketch predictions for the following 12 months.

g How confident do you feel about these predictions?

____

____

____

ISBN: 9780170370424

2 The graph below shows the total number of students absent each day from Paradise High School for the first seven weeks of a term.

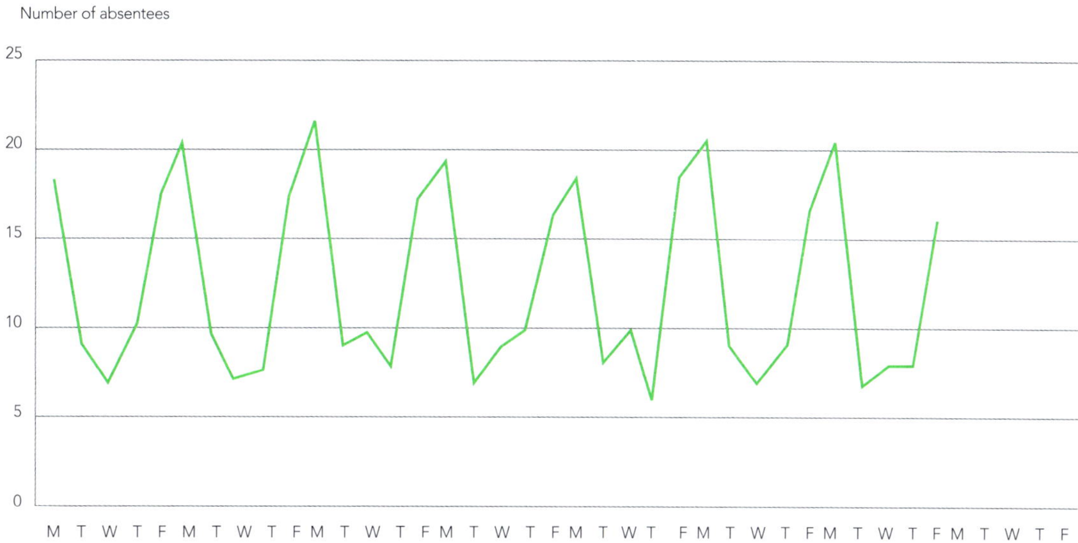

a Draw a trend line on the graph

b How many students were absent on the first Monday of this data?

____

c Which day had the fewest absentees, and how many were there?

____

d Describe any trends.

____

____

____

____

e Describe the repeating pattern.

____

____

____

____

____

f On the graph, sketch predictions for the following five school days.

g How confident do you feel about these predictions?

____

____

____

ISBN: 9780170370424

3 A website is launched at midnight on Sunday night. The numbers of hits each hour are recorded over the first three days.

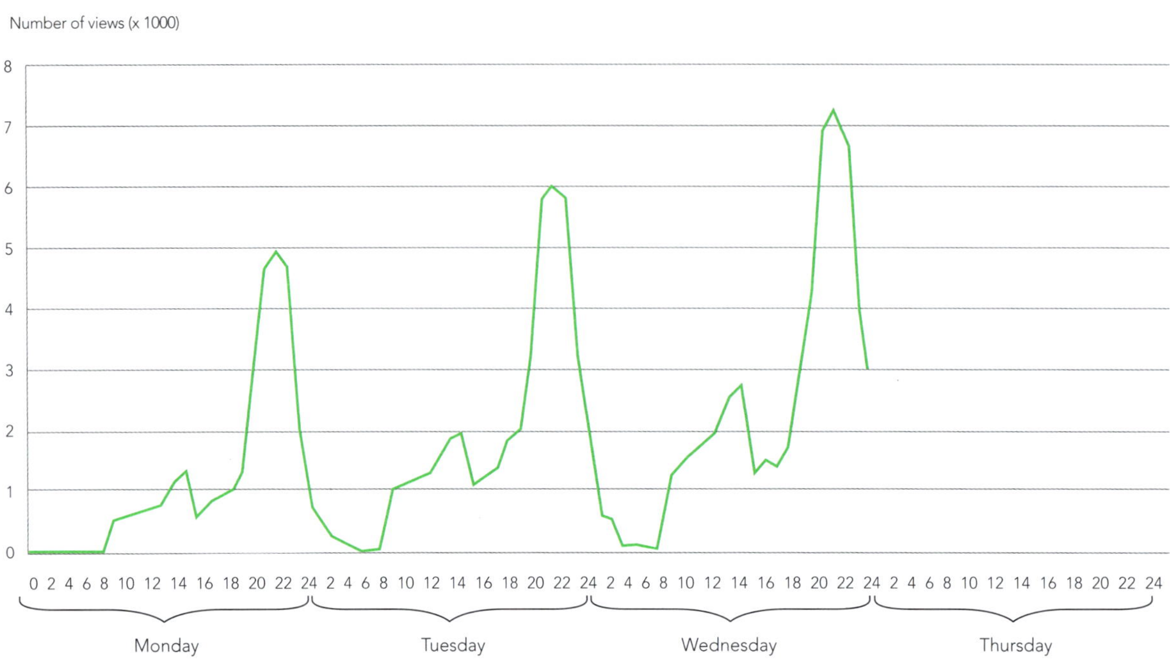

a Draw a trend line on the graph

b Estimate the highest number of hits per hour on Monday, and when did this occur?

____

c Describe any trends.

____

____

____

____

d Describe the repeating pattern.

____

____

____

____

____

e On the graph, sketch predictions for the next 24 hours.

f How confident do you feel about these predictions?

____

____

____

ISBN: 9780170370424

4 The total number of hits per day on the new website are also recorded, and are shown below.

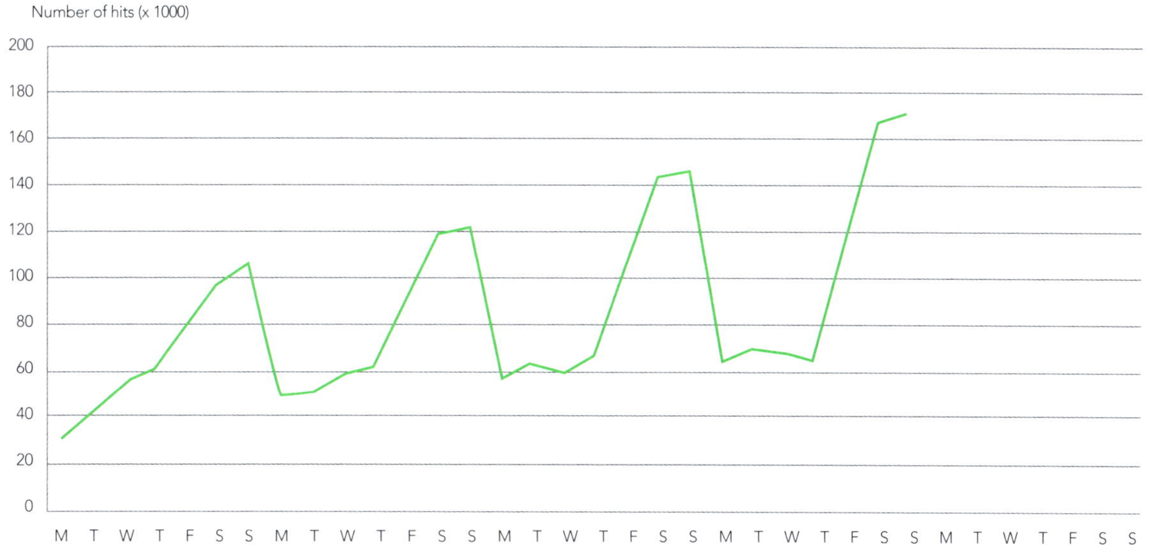

Number of hits (x 1000)

200
180
160
140
120
100
80
60
40
20
0

M T W T F S S M T W T F S S M T W T F S S M T W T F S S M T W T F S S

a Draw a trend line on the graph

b Estimate the number of hits on the first Sunday after the website is launched.

c Describe any trends.

d Describe the repeating pattern.

e On the graph, sketch predictions for the next week.

f How confident do you feel about these predictions?

ISBN: 9780170370424

5 The graph below shows the total rainfall (mm) each month for a South Island town.

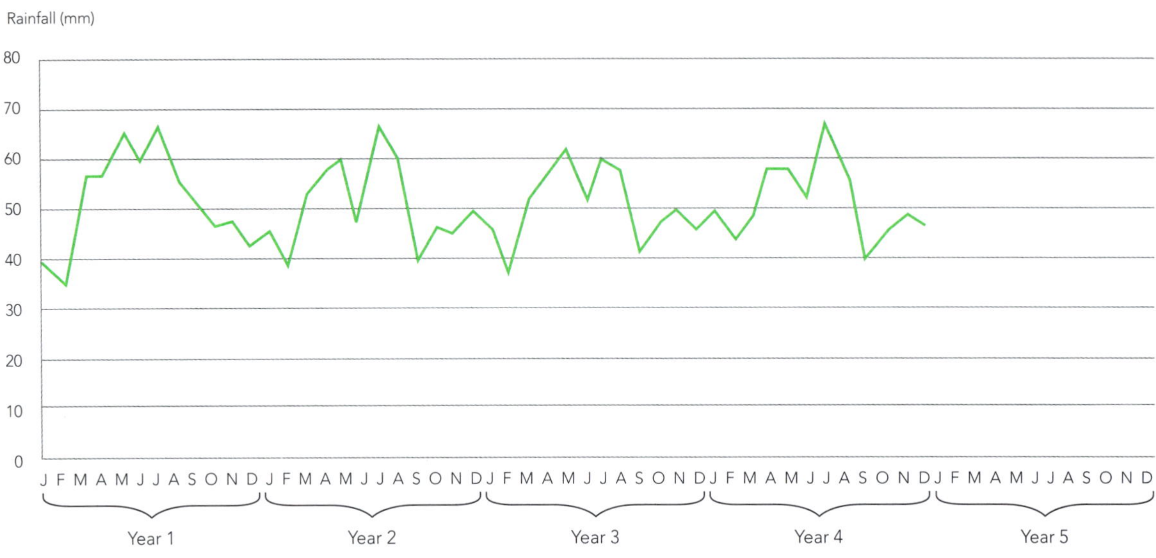

a Draw a trend line on the graph

b Estimate the rainfall during the February of Year 1.

c Describe any trends.

d Describe the repeating pattern.

e On the graph, sketch predictions for the next year.

f How confident do you feel about these predictions?

ISBN: 9780170370424

Comparing time series

Example: The graphs below show the rainfall for a South Island and a North island town.

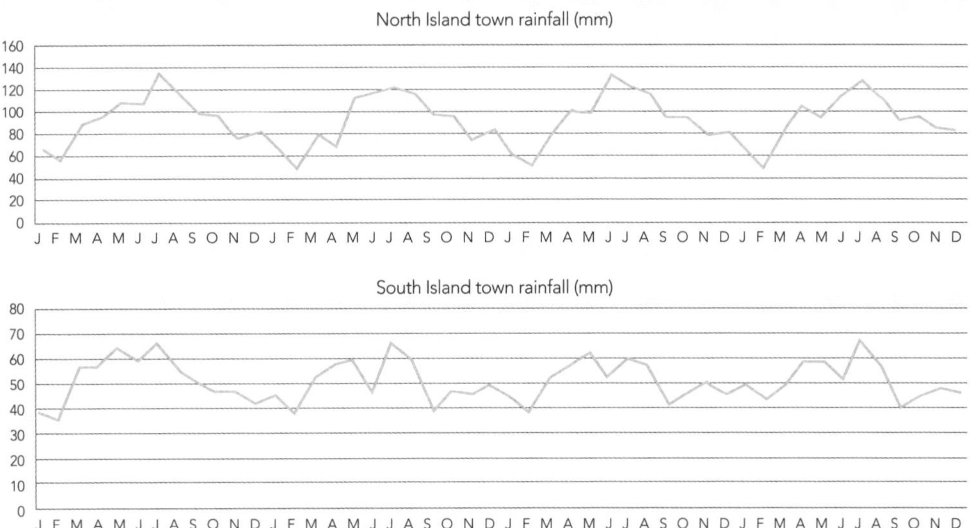

a Susie is considering living in one of these towns. She dislikes days when it rains all the time. Which town would you recommend that she selects? Justify your answer by referring to the graphs.

Probably she should choose the South Island town because the driest months have a rainfall of 35–40 mm, whereas the driest months in the North Island town get over 50 mm. The wettest months in the South Island town get 60–70 mm, but in the North Island town they get over 120 mm. However, these are monthly averages so we have no information about how the rain is distributed from day to day, or within days. It is possible that in the South Island town it drizzles for long periods, but the rain in the North Island town might come in deluges during the night.

b Wherever she selects, she wants to have a big garden that she doesn't need to water often, particularly when she goes away in summer. With this in mind, which town would you recommend that she selects? Justify your answer by referring to the graphs.

She should choose the North Island town because it gets a lot more rain overall. The time of day and the intensity, about which we know nothing, don't matter much to plants. In addition, the rainfall in the North Island town is much more predictable: it is much more regular with less variation. In the South Island town, February would be a problem time for a garden, because this is when it is usually warmest when gardens need most water. However, this is the driest time of the year, often with less than 40 mm over the whole month, so it would certainly need watering while she was away.

c Discuss the limitations of this data, or any other research she should do before making a decision.

She should find out how the rain is distributed throughout the day, and from day to day, because she only has monthly averages. She should also get information about other factors like temperature and wind conditions. She should study data for a longer period — this is only four years.

ISBN: 9780170370424

Answer the following questions.

1 The following graphs show average monthly temperatures for a North Island town and a South Island town.

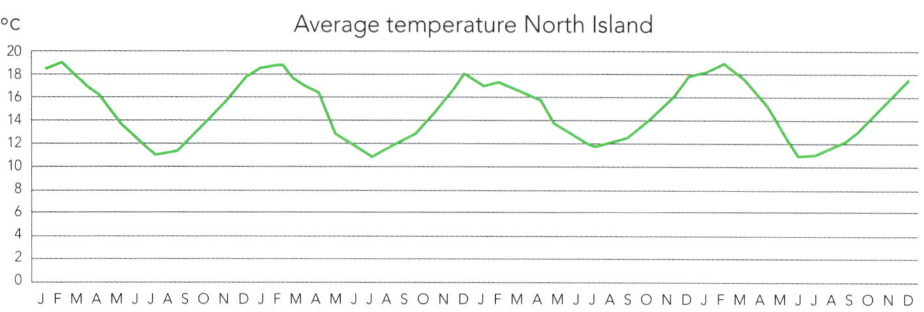

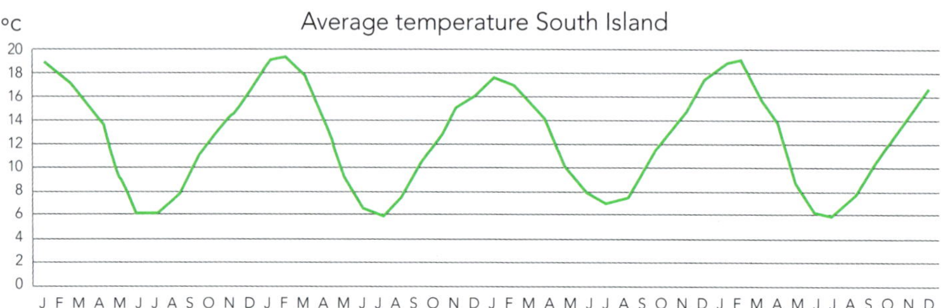

a Describe the similarities between the average monthly temperatures in the North Island town and the South Island town.

b What differences are there between the two towns?

c Susie likes to have a climate with cold overnight temperatures in winter, to kill off some of the garden pests. Which town would suit her better in this respect?

ISBN: 9780170370424

2 Brad and Abbie had 10 Algebra tests throughout each of Years 9 and 10. Their results have been plotted on the graph below. Each test was marked out of 20.

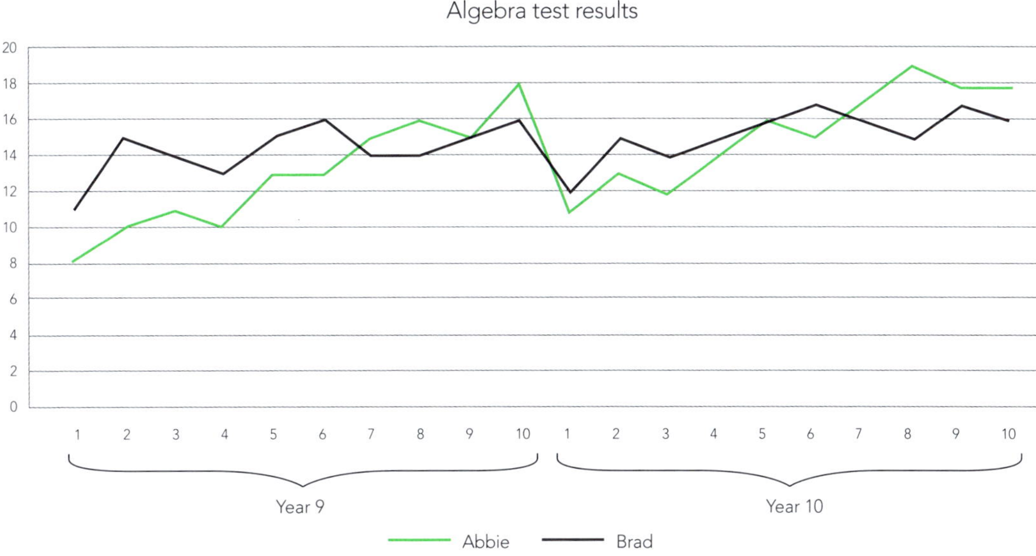

	Abbie	Brad
Minimum	8	11
Lower Quartile	13	14
Median	14	15
Upper Quartile	15	16
Maximum	18	17
Mean	14.1	14.8

a By referring to the graph and/or the table, discuss any similarities in their results.

b By referring to the graph and/or the table, discuss how their results differ.

c Brad maintains that he is better at Algebra. Do you agree? Justify your answer by referring to the graph and the table.

3 The school canteen decides that they have room in their fridges for either flavoured milk or juice. They currently sell both. They take up about the same amount of space, and the profit is the same on each. In order to decide which to keep, they record how many they sell each day throughout a term.

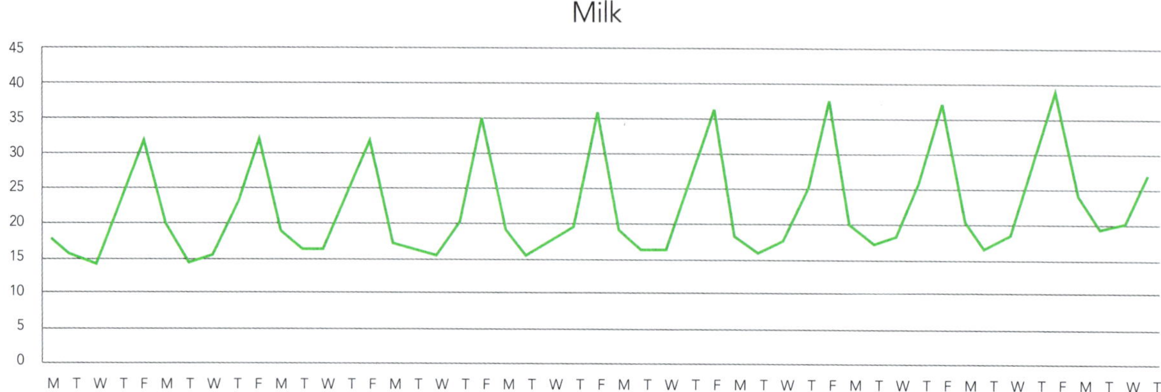

a Complete the table.

	Milk	Juice
Minimum	14	12
Lower Quartile	16	15
Median	19	19.5
Upper Quartile	26	29
Maximum	39	46
IQR		
Range		
Mean	22.28	24.16

ISBN: 9780170370424

b What similarities are there in the sales of milk and juice? Justify your answer by referring to the graphs and the table.

c Are there differences between the sales of milk and juice? Justify your answer by referring to the graphs and the table.

d From a **statistical** point of view, would you recommend that the canteen continues to stock milk or juice? Justify your answer.

e Discuss any limitations of the data, or other research needed before making a decision.

4 A roadside café records the values of coffee and ice cream sales for each quarter of the year: March, June, September and December. For example, the value given for the March quarter is the sum of the total sales for January, February and March.

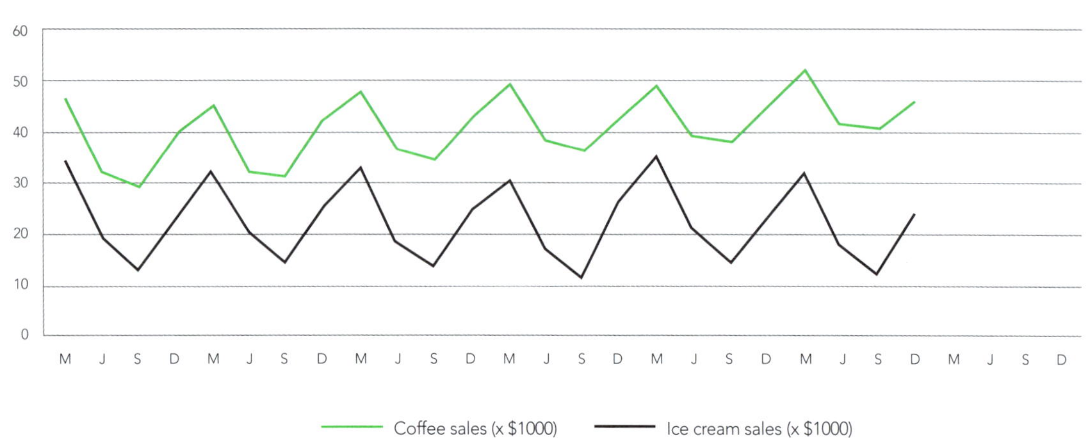

a What were the values of coffee and ice cream sold during the first September quarter?

____

b Draw trend lines on the graph for the sales of coffee and ice cream.

c What similarities are there in the sales of coffee and ice cream? Justify your answer by referring to the graphs.

____

____

____

____

d Are there differences between the sales of coffee and ice cream? Justify your answer by referring to the graphs.

____

____

____

____

e On the graph, sketch predictions for the next four quarters (one year).

f How confident do you feel about these predictions?

____

____

____

____

ISBN: 9780170370424

Statistical reports

Use what you have learnt in this book, along with Years 9 and 10 statistics to answer the following:

1 A local newspaper contains items about the city zoo. It published this graph showing the average number of visitors each month over the last five years along with the number of visitors each month during 2015.

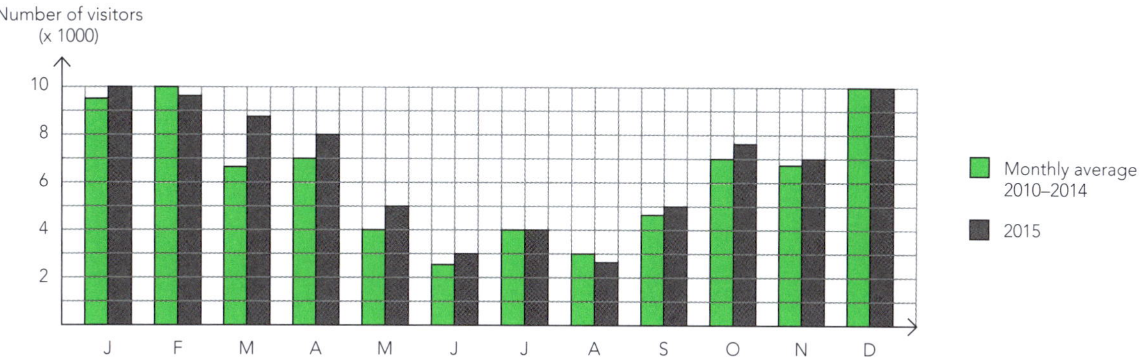

a Describe how the average monthly attendance changes throughout the year. Can you explain any irregularities in the attendance?

b During 2015 the zoo displayed gorillas for the first time. When was the gorilla enclosure opened to the public? Describe how it affected the number of visitors.

2 The newspaper included a quote from the manager stating "the annual number of visitors to the zoo has increased dramatically over the six years". He produced the following graph to support his statement:

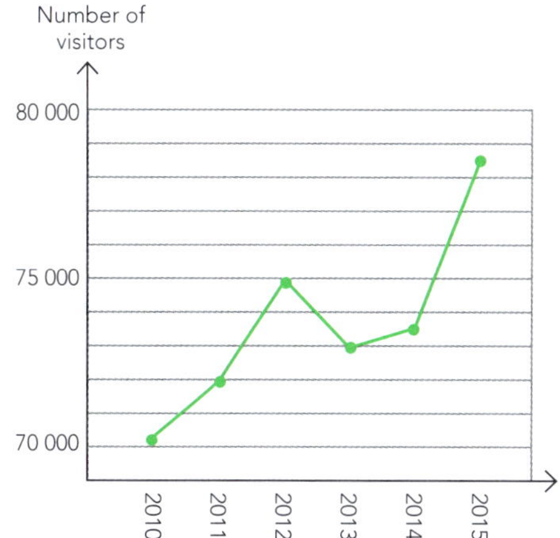

Do you agree with him? Justify your answer.

ISBN: 9780170370424

3 The number of visitors to the zoo for each day of the week during 2015 was reported.

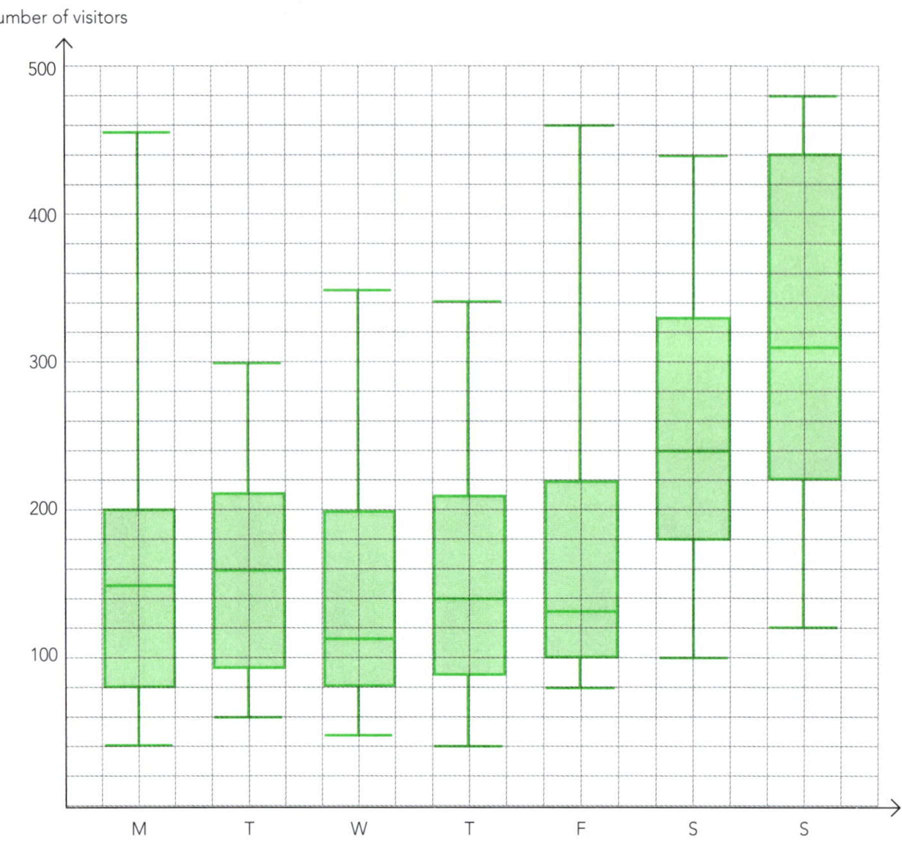

a Which day of the week usually gets the most visitors? ______________________________

b Which day of the week gets the biggest range for the number of visitors?

____

c Which day of the week has the greatest spread in the number of visitors?

____

Hint: Consider the **best** measure of spread.

d Which two days have the most skewed distributions of the number of visitors?

____

e The box plots for Mondays and Fridays differ from those for Tuesdays, Wednesdays and Thursdays. In what way do they differ, and why might this be?

____

____

4 "There was great excitement during the last month at the arrival of three new baby capuchin monkeys. This boosts the zoo's population to eight." However, the manager is concerned that the mean weight of the eight monkeys is now only 0.95 kg, when it should be 1.3 kg. Should the manager be concerned? Give statistical reasons for your answer.

____

____

____

 ISBN: 9780170370424

5 Over the last year, the manager kept records of the ages and, for the top three countries, where the visitors were from.

Countries	≤ 10	10 < 20	20 < 40	40 < 60	≥ 60	Totals
New Zealand	8 992	7 551	6 489	3 191	3 690	**29 913**
Australia	1 566	1 783	3 623	862	2 871	**10 705**
China	588	477	3 134	2 554	2 679	**9 432**
Other countries	1 301	1 466	4 467	2 477	4 012	**13 723**
Totals	**12 447**	**11 277**	**17 713**	**9 084**	**13 252**	**63 773**

a How many visitors were Australians who were at least 60 years old? ______________

b How many visitors were from Australia or New Zealand, and were under 10 years old or younger?

____

c What proportion of visitors was from New Zealand? ______________

d Calculate the probability that a visitor was between 20 and 40 years of age.

____

e Calculate the probability that a visitor was Chinese and 10 years old or younger.

____

f Calculate the probability that a Chinese visitor was 10 years old or younger.

____

g Which group contained the biggest proportion of children who were 10 years old or younger? Support your answer with calculations.

____

____

____

6 The newspaper printed a pie graph showing the percentage of the budget spent on different groups of animals.

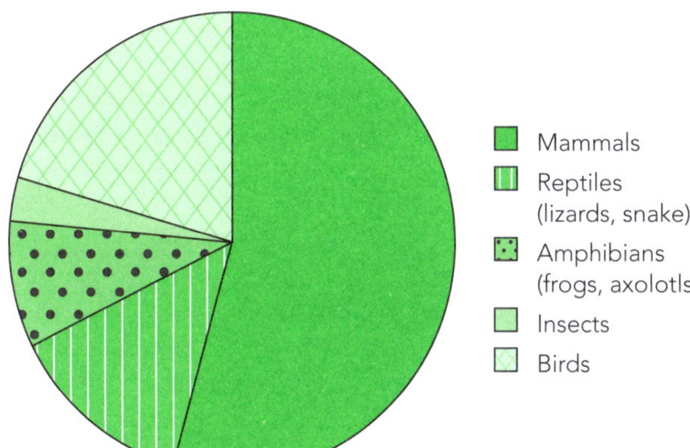

a About a quarter of the budget is spent on which three groups of animals?

____

____

b If the budget was $ 1.35 million, estimate the amount spent on the mammals.

____

____

ISBN: 9780170370424

Practice questions

1 Hemi and Tara are playing Snakes and Ladders. These are the rules:

- When it is your turn, you throw a die and move forward the number of squares shown.
- If you land on the tail of a snake (e.g. square 42), you slide down to a lower number.
- If you land on the base of a ladder (e.g. square 39 or 41), you climb to the top of the ladder.
- If you throw a six, then you move forward six squares, then throw the die again, and move forward the number of squares shown.

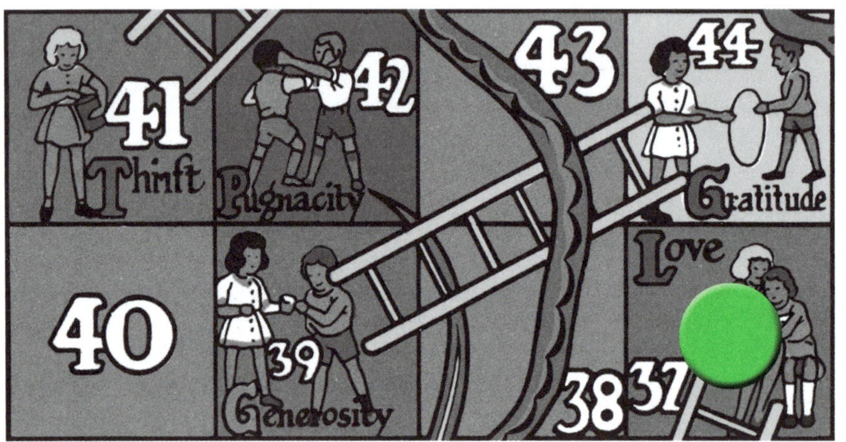

a Hemi is on square 37, and it is his turn. What is the probability that during this turn he will land on the tail of a snake?

b What is the probability that he lands on the base of a ladder during this turn?

c What is the probability that he lands on square 44 during this turn?

d Suppose Hemi is on 37 and Tara is on 38. Tara says that in the next two turns, she is more likely to land on the ladders on squares 39 and 41 because she is closer to them. Is she correct? Justify your decision.

 ISBN: 9780170370424

Hemi and Tara did a survey of the people in their class. They asked how many had played Snakes and Ladders, and how many had played Monopoly. The results are shown below.

	Played Monopoly	Had not played Monopoly
Played Snakes and Ladders	9	3
Had not played Snakes and Ladders	6	10

e What is the probability that a student chosen randomly from the class had played Monopoly?

f What is the probability that a student chosen randomly from the class had played both games?

g What is the probability that a Monopoly player had also played Snakes and Ladders?

h If two students were chosen randomly from the class, what is the probability that both played Snakes and Ladders but *not* Monopoly?

i Hemi and Tara would like to improve their survey of how many students at their school have played Monopoly or Snakes and Ladders. Suggest some ways in which they could do this, and explain your reasoning.

ISBN: 9780170370424

2 The shop at a service station recorded the values of different products sold each day. They drew a graph of the value of ice creams and the value of sunglasses sold each day.

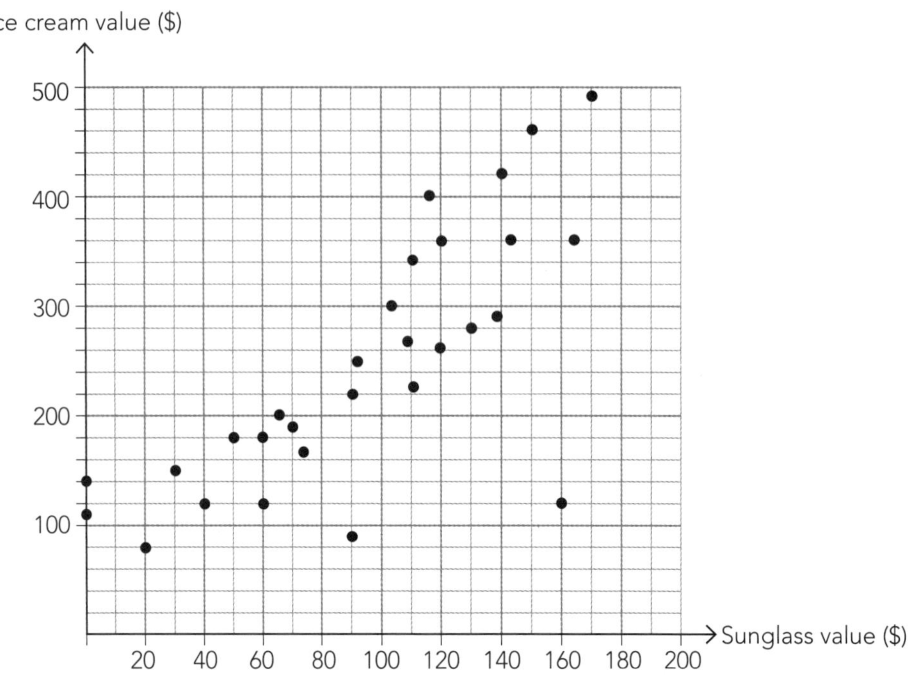

a Draw a line of best fit.

b On one day, the shop sold $150 worth of ice cream. What was the value of the sunglasses sold that day?

c On how many days did the shop sell less than $100 worth of sunglasses?

d Do you consider that there are any unusual points on the graph? If so, list the coordinates and explain why.

e Describe the relationship between sunglass and ice cream sales.

ISBN: 9780170370424

For 100 people who bought ice creams, they also recorded how many each person bought. The results are shown on the graph below.

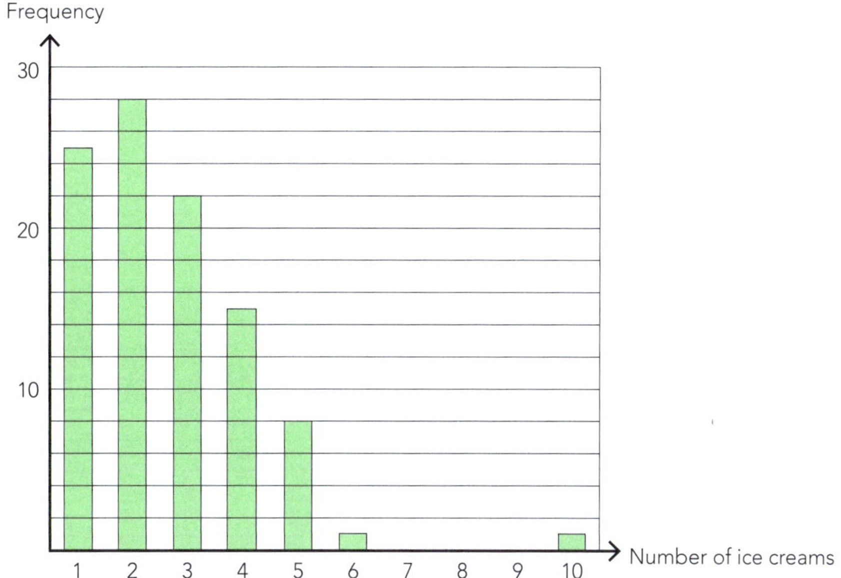

f What percentage of people bought three ice creams?

g How many people bought four or more ice creams?

h If 500 people bought ice creams from the service station, how many would you expect to buy four or five ice creams?

ISBN: 9780170370424

3 A scheme is trialled in which one class from School A and one class from School B does 10 Algebra tests at intervals throughout each of Years 9 and 10. Test 1–10 are done in Year 9 and tests 11–20 are done in Year 10. The average marks for each test are shown in the graph below, along with a box plot of the results for each class. All the tests are marked out of 20.

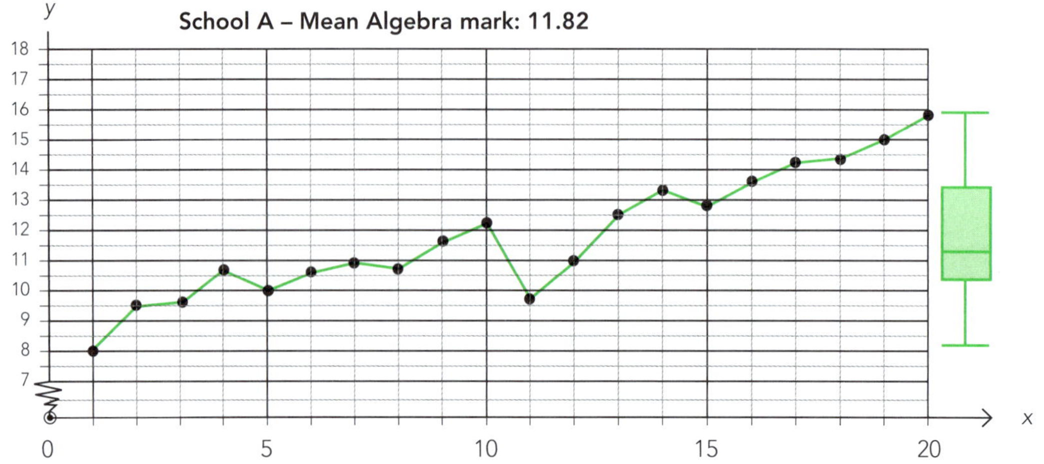

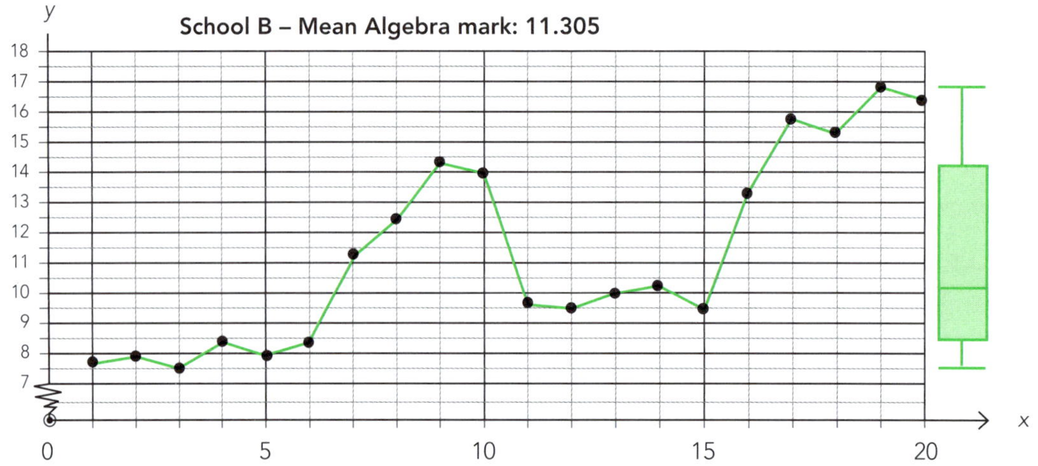

a What were the highest average marks obtained by each class, and when did these occur?

____

____

____

b What are the similarities between the two sets of marks? Justify your answer by referring to the graphs.

____

____

____

____

 ISBN: 9780170370424

c What are the differences between the two sets of marks? Justify your answer by referring to the graphs.

d Do the two graphs suggest that there are differences in the way in which Algebra is taught at the two schools? Justify your answer.

e Which school did better on average?

f Would you expect to come to the same conclusion if the same tests were done on all Year 9 and Year 10 students at the two schools?

g Overall, which method of teaching Algebra do you think is more effective?

h How could you improve this investigation?

Answers

Probabilities are rounded to a maximum of 4 dp. Professional judgement should apply.

Probability (pp. 7–28)

The range of values for probabilities (p. 7)

0	impossible, no chance, no way
0.1	very unlikely, slight chance, possible
0.2	improbable, possible, unlikely
0.3	improbable, maybe, possible, unlikely
0.4	maybe, possible, unlikely
0.5	maybe, possible
0.6	likely, probable
0.7	likely, probable
0.8	very likely, probable
0.9	very likely, extremely likely
1.0	definite, a sure thing

Using numbers for writing probability (p. 8)

1 $\frac{7}{16} = 0.4375$ $\frac{1}{3} = 0.\dot{3}$ Most likely: $\frac{7}{16}$

2 $\frac{3}{4} = 0.75$ $\frac{8}{11} = 0.\dot{7}\dot{2}$ Most likely: $\frac{3}{4}$

3 $\frac{17}{18} = 0.9\dot{4}$ $\frac{19}{20} = 0.95$ Most likely: $\frac{19}{20}$

4 $\frac{13}{24} = 0.541\dot{6}$ $\frac{27}{50} = 0.54$ Most likely: $\frac{13}{24}$

5 $\frac{11}{20} = 0.55$ $\frac{7}{13} = 0.5385$ Most likely: $\frac{11}{20}$

6 $\frac{6}{17} = 0.3529$ $\frac{3}{8} = 0.375$ Most likely: $\frac{3}{8}$

Test yourself

1 ✔ 0.009 is between 0 and 1, so it is a valid probability.

2 ✖ -0.35 is negative, and negative probabilities do not exist.

3 ✖ 105% = 1.05. Probabilities greater than 1 are not possible.

4 ✔ $\frac{1}{200}$= 0.005 which is between 0 and 1, so it is a valid probability.

Ways of calculating probabilities (pp. 9–13)

1 Equally likely outcomes (pp. 9–10)

1 $0.1\dot{6}$ **2** 0.5
3 0.5 **4** $0.8\dot{3}$
5 0 **6** 1
7 0.25 **8** 0.5
9 0.0769 **10** 0.0385
11 0.1538 **12** 0.1154
13 0.9231 **14** 0.8462
15 $0.291\dot{6}$ **16** $0.708\dot{3}$
17 $0.41\dot{6}$ **18** $0.58\dot{3}$
19 $0.791\dot{6}$ **20** $0.791\dot{6}$
21 0 **22** 1
23 36 **24** (1 2), (2 1)
25 $0.0\dot{5}$ **26** $0.\dot{1}$
27 $0.1\dot{6}$ **28** $0.08\dot{3}$
29 $0.91\dot{6}$ **30** 0.5
31 $0.1\dot{6}$ **32** $0.41\dot{6}$

33

	1	2	3	4	5	6
1	1 1	1 2	1 3	1 4	1 5	1 6
2	2 1	2 2	2 3	2 4	2 5	2 6
4	4 1	4 2	4 3	4 4	4 5	4 6
8	8 1	8 2	8 3	8 4	8 5	8 6

34 $0.041\dot{6}$ **35** 0.125
36 0.5 **37** $0.1\dot{6}$

2 Long run relative frequency (pp. 11–13)

1 P(bus) = 0.7
P(walked) = $0.22\dot{7}$
P(driven) = $0.07\dot{2}$

2 P(Excellence) = 0.25
P(Merit) = 0.4643
P(Achieved) = 0.2143
P(Not achieved) = 0.0714

3 **a** 0.225 **b** 0.5938 **c** 40.63%
4 **a** 0.0813 **b** 0.4563 **c** 91.88%
5 **a** 39 **b** 0.325 **c** 0.65

6 **a**

Trial #	Result	Total 'downs'	% of drops	P ('down')
1	down	1	100	1.0
2	up	1	50	0.5
3	down	2	67	0.67
4	down	3	75	0.75
5	up	3	60	0.6
6	down	4	67	0.67
7	up	4	57	0.57
8	up	4	50	0.5
9	up	4	44	0.44
10	down	5	50	0.5
11	down	6	55	0.55
12	down	7	58	0.58
13	up	7	54	0.54
14	down	8	57	0.57
15	up	8	53	0.53

Table continued over page.

ISBN: 9780170370424

Trial #	Result	Total 'downs'	% of drops	P ('down')
16	down	9	56	0.56
17	down	10	59	0.59
18	up	10	56	0.56
19	down	11	58	0.58
20	down	12	60	0.60

b

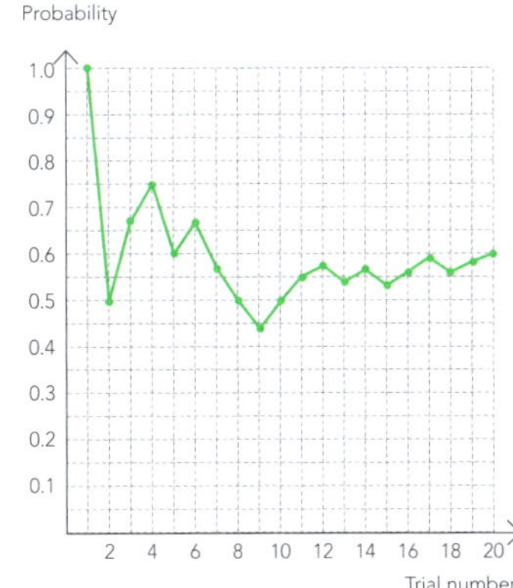

c Between 0.5 and 0.6 or 0.7, between 0.55 and 0.65, or any other reasonable **range** of values.

d Not very certain — she did only 20 trials and to be more certain she would need to do more. Or reasonably certain — the graph has been between 0.5 and 0.6 for the last 10 trials.

Expected number of outcomes (p. 14)

1 40 heads **2** 30 times
3 79 or 80 people **4** 30 or 31 boys
5 1 or 2 eggs **6** 1 hand

Probabilities from bar graphs and frequency histograms (pp. 16–18)

1 **a** 60 **b** $0.6\dot{3}$
  c 65% **d** 98
2 **a** 18 **b** $0.2\dot{7}$
  c $0.\dot{6}$ **d** 40
3 **a** 0.18 **b** 0.05
  c 7% **d** 1250 or 1251
4 **a** 0.04 **b** 0.73
  c 0.09 **d** 138
5 **a** 90 **b** $0.1\dot{4}$
  c $0.1\dot{6}$
  d She only tossed the die 90 times, so by chance she may have got fewer sixes than she should. If she had performed the experiment a lot more times, then, provided it was a fair die, she could expect a probability of closer to $0.1\dot{6}$.
  e 20
  f It would be similar with most bars between 12 and 18 units high, but which die numbers have the highest and lowest bars cannot be predicted.
  g The heights of the bars would be more similar, and about 167 units high.

Combining probabilities — probability trees (pp. 21–25)

1 **a**

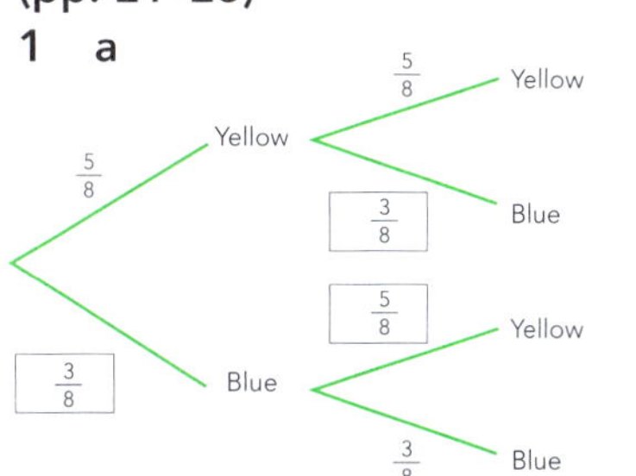

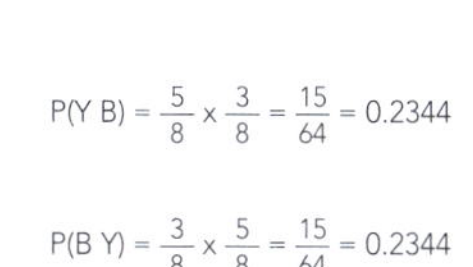

b $\frac{25}{64}$ or 0.3906 **c** $\frac{30}{64}$ or 0.4688
d $\frac{34}{64}$ or 0.5313 **e** $\frac{9}{64}$ or 0.1406
f $\frac{39}{64}$ or 0.6094 **g** 78 or 79
h 121 or 122

2 **a**

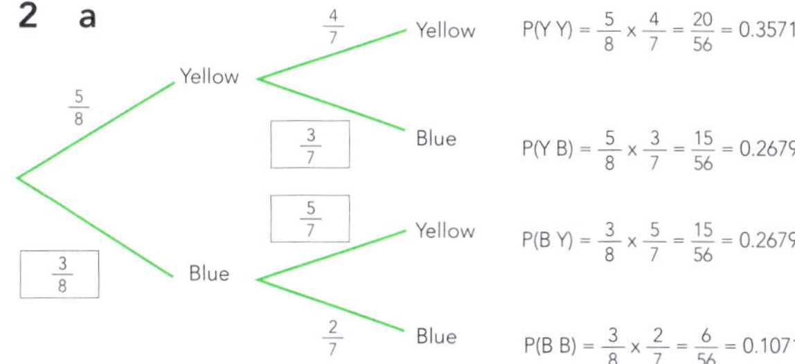

b $\frac{20}{56}$ or 0.3571 **c** $\frac{30}{56}$ or 0.5357
d $\frac{26}{56}$ or 0.4643 **e** $\frac{6}{56}$ or 0.1071
f $\frac{36}{56}$ or 0.6429 **g** $\frac{30}{56}$ or 0.5357
h 1

3 **a**

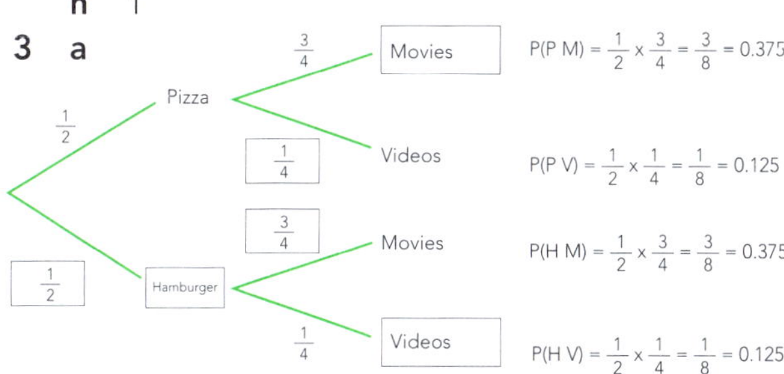

b $\frac{6}{8}$ or 0.75 **c** $\frac{1}{8}$ or 0.125
d $\frac{3}{8}$ or 0.375 **e** $\frac{7}{8}$ or 0.875
f $\frac{5}{8}$ or 0.625 **g** $\frac{3}{8}$ or 0.375
h $\frac{1}{8}$ or 0.125

ISBN: 9780170370424

4 a

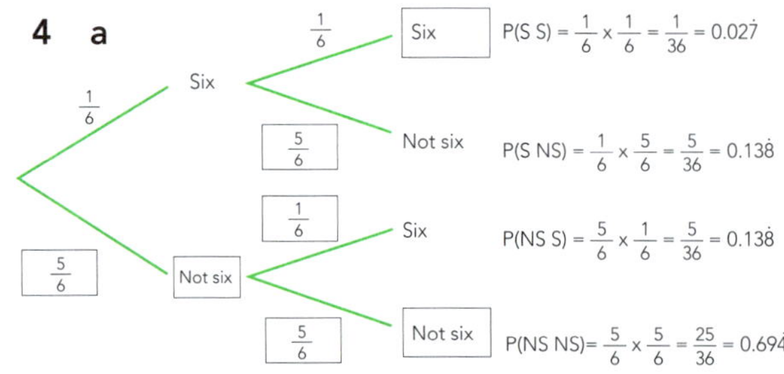

b $\frac{1}{6}$ or $0.01\dot{6}$ c $\frac{5}{36}$ or $0.13\dot{8}$

d $\frac{11}{36}$ or $0.30\dot{5}$ e $\frac{25}{36}$ or $0.69\dot{4}$

f They add to one because you must either get a six or not get a six in the first two throws.

g 24 or 25

h $\frac{25}{36} \times \frac{1}{6} = \frac{25}{216}$ or 0.1157

5 a $\frac{1}{4}$ or 0.25 b $\frac{1}{2}$ or 0.5

c 75% d 25

6 a $\frac{24}{100}$ or 0.24 b 62%

c 246 or 247 d 115 or 116

Two-way frequency tables (pp. 27–29)

1 a 79.6% b $\frac{87}{225}$ or $0.38\dot{6}$

c $\frac{87}{129}$ or $0.67\dot{4}$ d $\frac{179}{225} \times \frac{178}{224} = 0.6322$

2 a $\frac{390}{939}$ or 0.4153 b $\frac{162}{316}$ or 0.5127

c $\frac{387}{623}$ or 0.6212 d $\frac{390}{939} \times \frac{389}{938} = 0.1722$

3 a $\frac{27}{84}$ or 0.3214 b $\frac{11}{84}$ or 0.1310

c $\frac{16}{45}$ or $0.3\dot{5}$ d $\frac{27}{84} \times \frac{26}{83} = 0.1007$

4 a 0.57 x 1200 = 684 b 0.18 x 1200 = 216

c $\frac{216}{540} = 0.4$ d 984

5 a 0.19 x 1200 = 228 b 0.44 x 1200 = 528

c $\frac{528}{660} = 0.8$ d 1 068

6 a $\frac{798}{1287} = 0.620$ b $\frac{119}{1287} = 0.0925$

c $\frac{119}{319} = 0.3730$ d $\frac{527}{699} = 0.7539$

e First class passengers had a much lower chance of dying (0.373) than passengers in general (0.620). Third class passengers were more likely to die (0.754) than passengers in general (0.620).

Data handling (pp. 30–45)

Collecting data (p. 31)

1 Just one class might not represent the whole of Year 10. It might be a top class, or a class that studies a particular group of subjects, e.g. Technology. So this would be biased.

2 The students who walk past the senior common room are more likely to be senior students, so junior students may be under-represented in the sample, so it would be biased and not reflect the whole school.

3 Students who are late, enter by another gate or are absent will not have a chance of being selected, so the sample would be biased.

4 This should give an unbiased sample of school students because all are equally likely to be selected.

5 This surveys only those who use the canteen, so presumably like the food, and can afford to buy it. Those who don't use the canteen will not be surveyed. It will not get the views of those who buy from the canteen at times other than lunch time.

6 The survey will be self-selected. Only those who are interested enough, have the time and can be bothered will respond.

7 This might produce a random sample, but it might result in more or fewer students of particular nationalities being selected, e.g. the letter L could favour Korean and Chinese students.

8 Because the students are selected randomly, this should produce an unbiased sample. However, if there are significantly different numbers of students in each level, the levels with smaller numbers might be over-represented.

Measures of centre (p. 32)

1 Mean = 2.6
Median = 2.5
Mode = 1
The mean or median because they are close to each other and there are no extremely big or small values. This means both are good measures of centre.

2 Mean = 17
Median = 18.5
Mode = 20
The best is the median because it is near the centre of the data. The mode is much too high. The mean is too low because it has been dragged down by the 2, which is much smaller than any other value.

ISBN: 9780170370424

Measures of spread (pp. 33–34)

1 Minimum = 5
Lower quartile = 11
Median = 17
Upper quartile = 21
Maximum = 25
Range = 20
Inter-quartile Range = 10

2 Minimum = 1
Lower quartile = 6.5
Median = 10
Upper quartile = 14
Maximum = 19
Range = 18
Inter-quartile Range = 7.5

3 Minimum = 0
Lower quartile = 1
Median = 2.5
Upper quartile = 5
Maximum = 8
Range = 8
Inter-quartile Range = 4

4 Minimum = 1
Lower quartile = 5
Median = 8
Upper quartile = 14
Maximum = 21
Range = 20
Inter-quartile Range = 9

5 Minimum = 7
Lower quartile = 10
Median = 12.5
Upper quartile = 15.5
Maximum = 21
Range = 14
Inter-quartile Range = 5.5

6 Minimum = 13
Lower quartile = 14.5
Median = 17
Upper quartile = 19.5
Maximum = 21
Range = 8
Inter-quartile Range = 5

Displaying data (pp. 35–37)

1 Dot plots (p. 35)

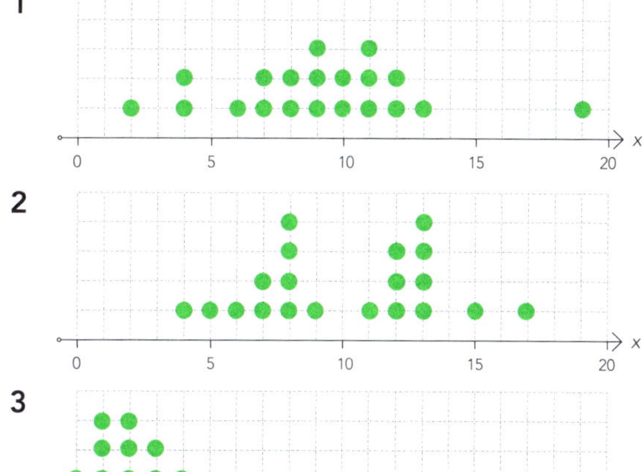

2 Box plots (p. 37)

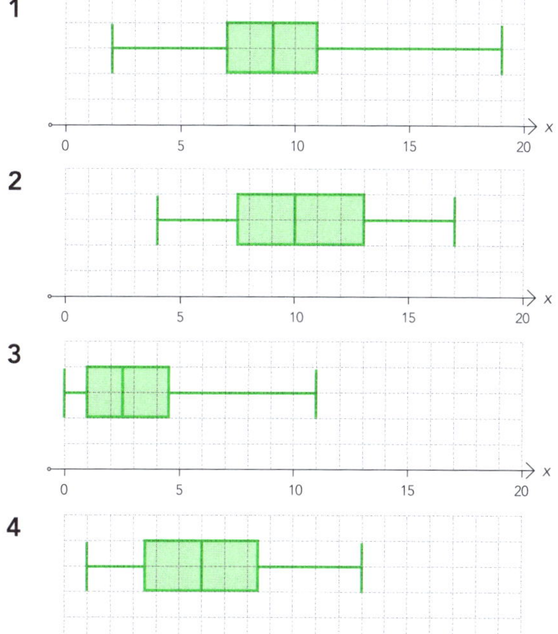

Combining dot plots, box plots and means (p. 38)

1 = b Mean is **about the same** as the median.
2 = c Mean is **below** the median.
3 = a Mean is **about the same** as the median.

Drawing conclusions from pairs of box plots (p. 40)

1 The boxes overlap and both medians are within the box of the other sample. Therefore we cannot conclude that the individuals in populations A will tend to be taller than those in B.

ISBN: 9780170370424

2 The boxes of the two samples overlap, but the median of sample B lies outside the box of sample A. We can conclude that individuals in population A are likely, on average, to be taller than those in population B.

3 The boxes do not overlap, and both medians lie outside the box for the other sample. Therefore, we can conclude that individuals in population B will tend to be taller than those in population A.

Comparing groups using dot and box plots (pp. 42–45)

1 a

	Algebra	Geometry
Minimum	5	4
Lower quartile	7	8
Median	9	10
Upper quartile	11	11.5
Maximum	15	15
Mean	9.52	9.65

b 9.65 – 9.52 = 0.13

c The students did slightly better on average in the Geometry test because the box for Geometry is a little further to the right, and the median for Geometry (10) is one more than that for Algebra (9). The means are very close, but only because two students got 15 and one got 14 in the Algebra test, and these dragged the mean up. There were also no very low marks in Algebra.

d No. Both of the medians are within the box for the other subject and the boxes overlap, so we cannot conclude that there would be a difference for the whole group. Also, this is data from just one class so it may not reflect the results of the whole of Year 11. This class might have been taught differently, results will vary with different teachers, etc.

e Randomly select a few students from each class, or get a roll of all Year 11 students and select every (say) tenth student.

2 a

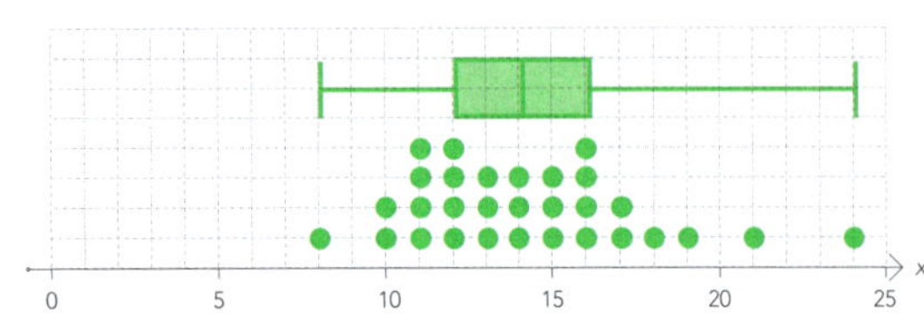

b Both the means (Boys 14.5 and Girls 14.2) and medians (Boys 15 and Girls 14) suggest that girls were slightly faster on average. However, there were more boys who could test very quickly. Seven boys took less than 10 seconds, while only one girl could do it in that time.

c Because both medians lie within the box of the other group and the boxes overlap, we cannot conclude that girls can text faster than boys. It is also possible that the sample was biased because it was taken outside the library. For example, this might be where the boys who play computer games go at lunch time, and playing computer games might make them faster at texting.

d Boys: IQR = 7 Girls: IQR = 4
This tells me that the times for boys were much more spread than the times for girls. This is partly because the boys' distribution is bimodal, with a group of boys who are very fast, taking 7–10 seconds, and a bigger group who took 14–20 seconds.

e Take bigger samples.
Make the selection of subjects freer of bias, by getting a roll of all Year 11 students and asking every (say) tenth student to answer the survey.

3 a

	Katherine Mansfield	Jane Austen
Minimum	4	3
Lower quartile	11	18
Median	15	28
Upper quartile	27	41
Maximum	45	59
Mean	18.5	29.2

b 28 – 15 = 13

c Both the medians (Jane Austen 28 and Katherine Mansfield 15) and the means (Jane Austen 29.2 and Katherine Mansfield 18.5) suggest that on average, Jane Austen wrote much longer sentences than Katherine Mansfield.

d Possibly. Both of the medians lie outside the other's boxes but the boxes overlap. There is evidence to suggest that Jane Austen's sentences are likely, on average, to be longer than Katherine Mansfield's.

ISBN: 9780170370424

- **e** The mean sentence length of Katherine Mansfield is 18.5, while the median is only 15. This is because the lengths of her sentences are skewed to the right. She wrote small numbers of very long sentences, which affected the size of the mean, but not the median.
- **f** Select more sentences. Select sentences from several different books.

4
- **a** 5.06 – 4.67 = 0.39.
- **b** The medians are the same, which suggests that there is no difference between their word lengths. However, the mean word length for Charles Dickens is slightly longer (0.36) than that for Jane Austen. This is because the distribution of the lengths of Charles Dickens' words is bimodal: he uses words with 10–12 letters, which Jane Austen doesn't. These have dragged the mean up, but did not affect the median. A bigger sample might be needed to clarify this.
- **c** Yes, because both medians lie within the other's box, and the boxes overlap. We cannot conclude that there is a difference between the lengths of words used.
- **d** Jane Austen's words were most consistent because her IQR is smaller (4) than Dickens' (5) and her range is much smaller (7) compared with Dickens', which is 11.
- **e** Take bigger samples of words. Take random samples of words from all their books, not just one.

Bivariate data (pp. 46–54)

1
- **a**

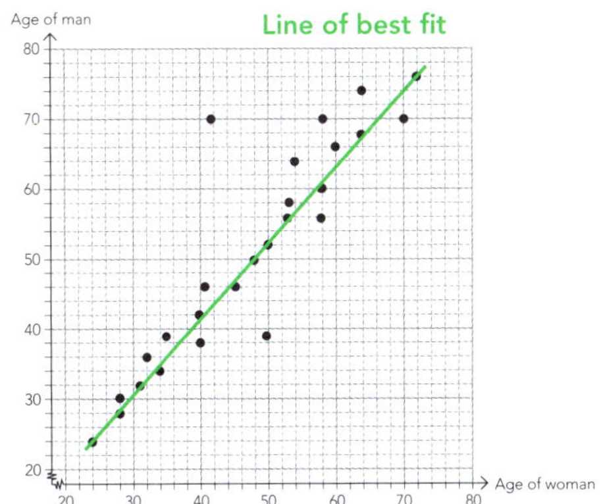

- **b** The man is 76.
- **c** Nine couples.
- **d** (42, 70) — the man was 70 and the woman 42, so the man was much older than the woman. (50, 39) — the man 39 and the woman 50. There are only three couples where the man is younger than the woman, and for the others the age difference was much smaller.
- **e**

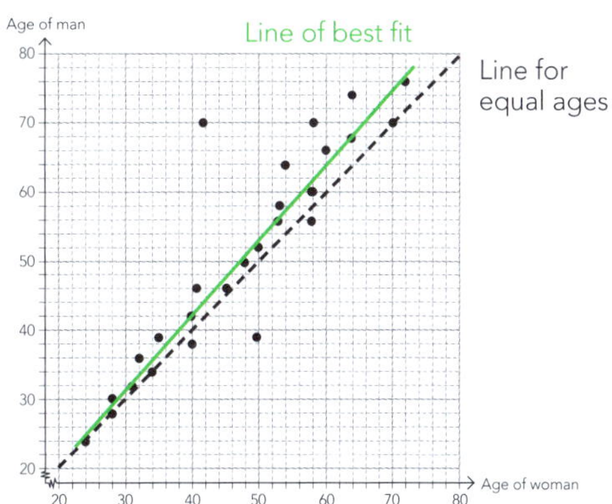

- **f** (58, 56) — woman 58, man 56
  (50, 39) — woman 50, man 39
  (40, 38) — woman 40, man 38
- **g** There is a positive and quite strong relationship between the ages of men and women in couples. It is strong because most points are close to the line of best fit. It is positive because young men coupled with young women, and older men coupled with older women. The age of the woman in a couple is usually a little less than the age of the man.
- **h** Because for each couple there are two pieces of numeric data — the age of the man and the age of the woman.

2
- **a**

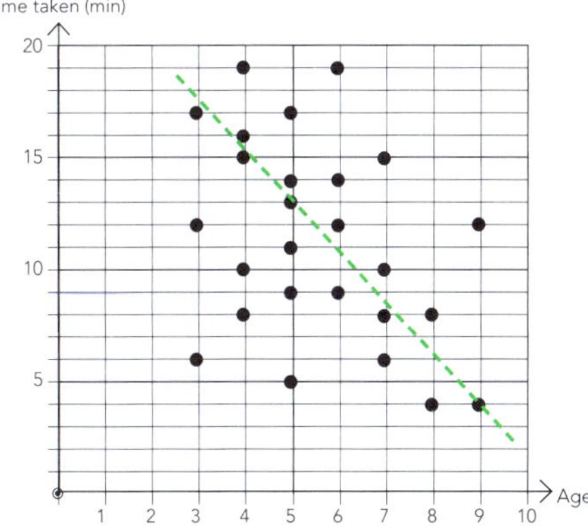

- **b** One took 4 minutes and the other took 8 minutes.
- **c** Two children.
- **d** There are no points on this graph that are a significant distance from other points. This is partly because all the points are quite scattered. So there are no unusual points.

e There is a weak negative relationship between children's ages and how long they took to solve the puzzle. It is negative because younger children took more time and older children took less time to solve the puzzle. It is weak because lots of points are a long way from the line of best fit — there is a lot of scatter. Some very young children did it very quickly and some older children were slow.

f I would time how long a lot more children took to do the puzzle.

3 **a**

% glass recycled

Line of best fit

100 90 80 70 60 50 40 30 20 10

10 20 30 40 50 60 70 80 90 100 % paper recycled

b Four countries.

c 64%

d (34, 64) and (40, 72) — both these points are a long way from other points and the line of best fit. They both recycle a lot more glass than paper. (The second of these points is New Zealand!)

e

% glass recycled

Equal glass and paper recycling

100 90 80 70 60 50 40 30 20 10

10 20 30 40 50 60 70 80 90 100 % paper recycled

f They recycle more paper packaging. This is shown on the graph; most points are to the right of the line for equal glass and paper recycling.

g The relationship between the recycling of paper packaging and glass packaging is fairly weak and positive. It is fairly weak because quite a lot of points are some distance from the line of best fit. It is positive because in general, countries that recycle a lot of paper also recycle a lot of glass.

4 **a**

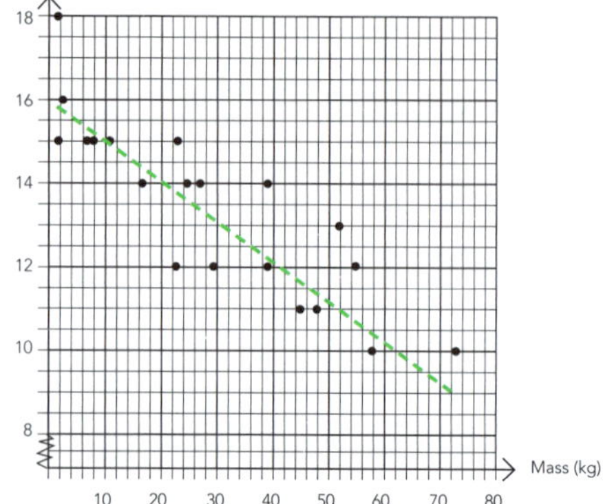

b 13 years

c 45 and 48 kg

d Because for each breed there are two pieces of numeric data: maximum lifespan and average mass.

e (2, 18) because it is a long way from the line of best fit, and because this breed has a maximum lifespan of two years longer than any other breed. (It's chihuahuas.)

f This relationship is moderately strong and negative. It is negative because heavier dogs have shorter lifespan, and lighter dogs live longer. The relationship is moderately strong because most points are reasonably close to the line of best fit.

5 **a**

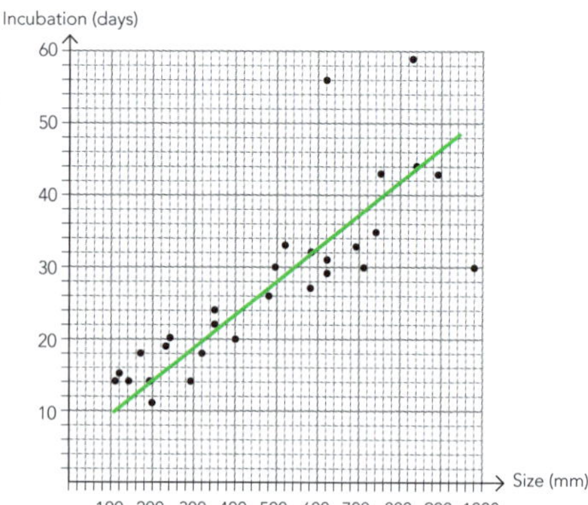

b 27 days

c 230 mm

d (980, 30) — a long way to the right of the line of best fit. Other birds with incubation periods of 30 days are between 500 and 700 mm.

(620, 56) and (830, 59) — both well above the rest of the data. The nearest point is a bird with an incubation period of 44 days.

 ISBN: 9780170370424

e There is a strong positive relationship between the size of a bird and the length of its incubation period. It is strong because most points are close to the line of best fit. It is positive because the bigger the bird, the longer its incubation period.

f Plot points for more species of birds.

Time series (pp. 54–66)

Single time series (pp. 56–60)

1 a

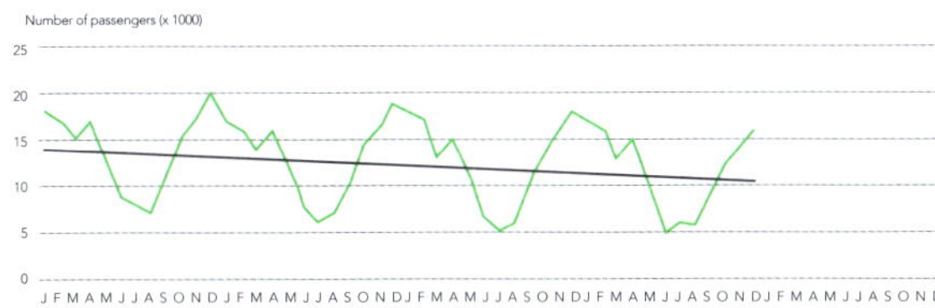

b 17 000 passengers

c 15 000 passengers

d There is a steady decrease of about 1000 passengers a year.

e The number of passengers is highest in December (18 000–20 000), after which the numbers drop each month until April, when there are about 2000 more passengers than would be otherwise expected. After that, the numbers drop to their lowest in June, July and August (5000–8000). Then there is a steady increase until they reach their peak in December again.

f

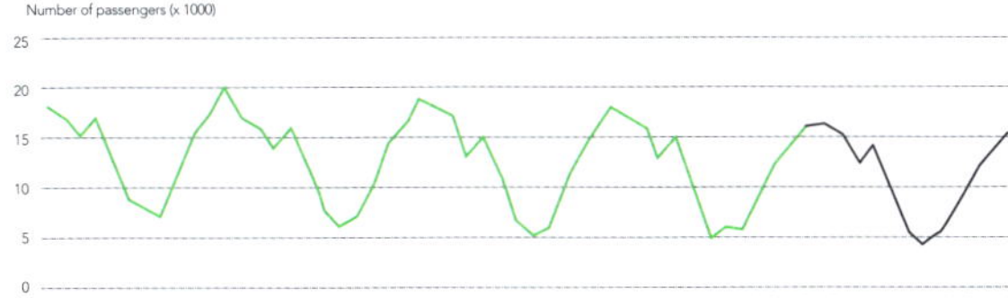

g Fairly confident because there is a reasonably regular decrease in both the peak values in December, and the lowest values in June, July and August. However, there is only four years of data, which is not a lot. In addition, many other factors may affect the numbers of people using the ferry — price of other modes of transport, economic factors, etc.

2 a

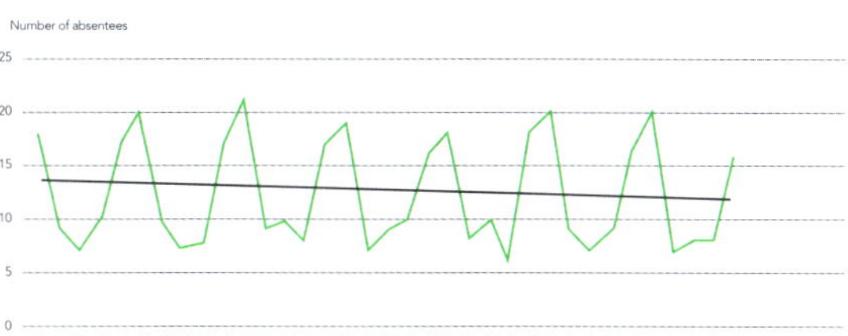

b 18 students

c The fifth Thursday and there were six students absent.

d The trend is flat —some weeks have more or fewer absentees, but the number doesn't increase or decrease over all.

e The highest number of absentees is on Mondays (17–21). The second highest is on Fridays (16–18). For Tuesdays, Wednesdays and Thursdays, the numbers of absentees is much lower than on Mondays and Fridays (6–10). There does not appear to be a pattern in the Tuesday to Thursday absences.

f

Number of absentees

g Fairly confident because the pattern is quite regular and there is seven weeks of data. However, there are other factors that might affect absences such as flu epidemics, students getting tired towards the end of term, etc.

3 a

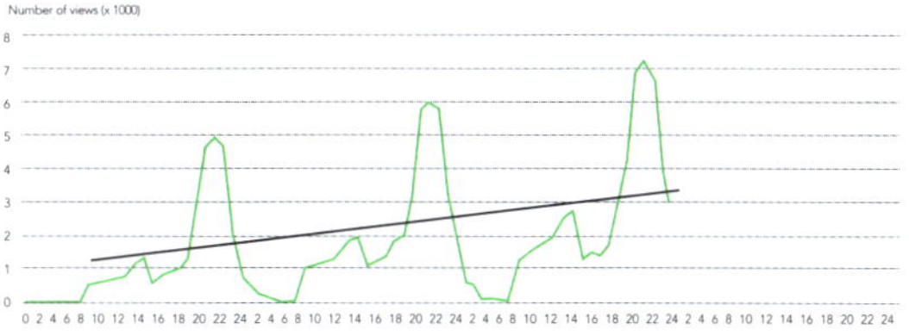

b 4900 hits at about 9 or 10 p.m.

c Apart from in the early hours of the morning, the number of hits per hour increases over the first three days.
The difference between the lowest number each day and the highest number in each day increases by about 1000 hits per day.
The evening peaks (8–10 p.m.) increase by about 1000 hits per day.

The lunchtime peaks (midday – 1 p.m.) increase by about 700 per day.
The lowest point on the graph (5–6 a.m.) when almost nobody is using the website, hardly change.

d During the few hours before 6 a.m., almost nobody uses the site. The number of hits increases steeply until about 8 a.m., then steadily until a peak at lunch time (midday – 1 p.m.). There is a steep drop-off during the afternoon, followed by a steep increase to the highest peak of the day in the evening between 8 and 10 p.m. There is a dramatic drop in use after midnight until about 1 p.m., then a steady decline to almost no use again between 4 and 6 a.m.

e

f The pattern was fairly consistent, so that makes me feel reasonably confident. However, there were only three cycles to base the prediction on, which is not many. There could also be unforeseen factors such as the crashing of the website, it might go viral, etc.

4 a

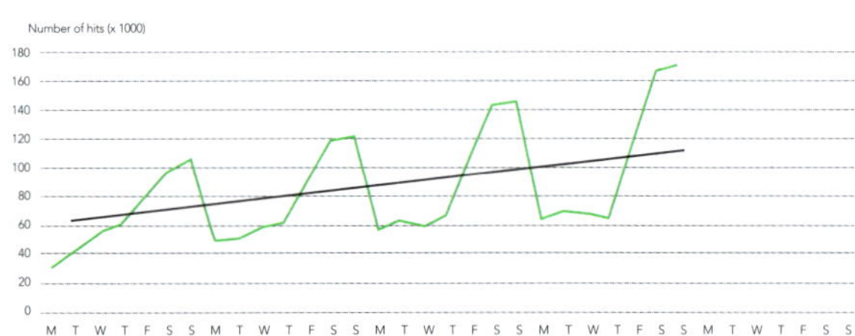

b 103 000–107 000 hits

c Overall the number of hits on the website is increasing. The weekend levels increase by about 20 000 each week. However, the number of hits on Monday to Thursday only increases by about 4000 to 5000. The difference between the number of hits on Mondays to Thursdays and the number of hits on weekend days is increasing each week. The drop from Sunday to Monday is about 55 000 in the first week, about 65 000 in the second week and almost 80 000 in the third week.

d The greatest number of hits is at the weekend, and the number of hits on Sundays is a little greater than the number of hits on a Saturday. Apart from the first week when the website was new, the number of hits on Mondays to Thursdays is much lower. There is an increased number of hits on Fridays, but not as many as on Saturday.

e

f The pattern was fairly consistent, so that makes me feel reasonably confident. However, there were only four cycles to base the prediction on, which is not many. There could also be unforeseen factors such as the crashing of the website, it might go viral, etc.

5 a

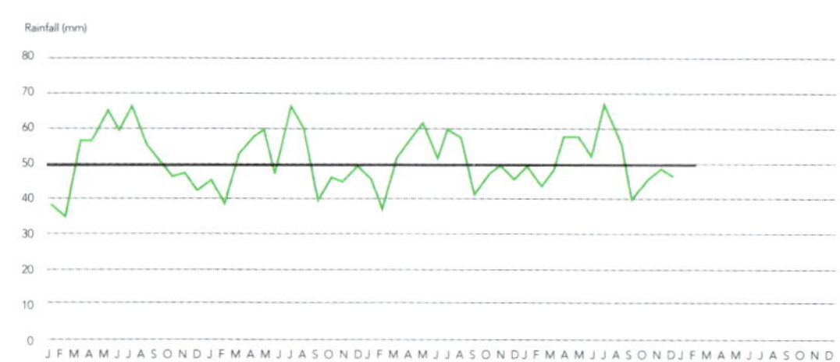

b 35–37 mm

c The rainfall is quite variable from month to month, but on average, it neither increases nor decreases over these four years.
The amount of variation in the monthly rainfall throughout the year is also similar over these four years.

d In general there is more rain during the winter months (April–August) than in summer (September–March). The wettest months are usually May or July, with June being a little drier. They usually have about 60 mm. However, this is variable. The driest month is often February, with 35–40 mm rainfall.

e

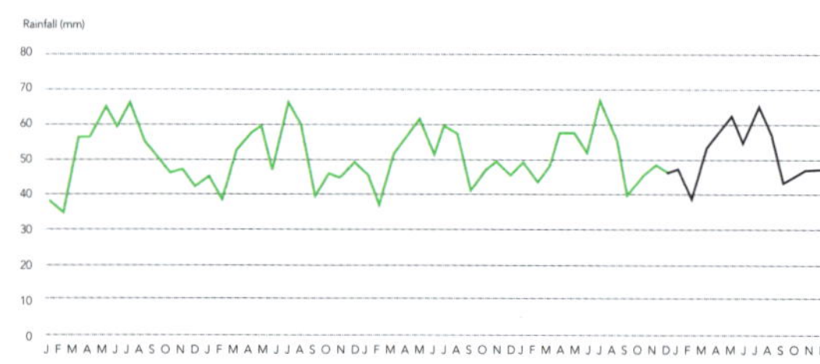

 ISBN: 9780170370424

f Not very confident because there was only four years' data and there is a lot of variability in it.

Comparing time series (pp. 62–66)

1 a The temperature in both towns follows a seasonal cycle, with high temperatures from December to March and low temperatures from May to August.
In both towns, the third summer was unusually cool: maximum of 18° in the north compared with about 19° in the other years. It was just over 16° in the south, compared with the usual 18°.
The third winter was warmer than the other years in both towns: it hardly dipped below 12° in the North Island compared with the usual 11°. In the South Island town the average did not dip below 6°, but in the other years it was about 5°.

b The maximum summer temperatures are fairly close (18° to 19°). However, the winter average temperatures are very different. In the North Island town they dip to averages of about 11° to 12°. In the South Island town they dip to 5° to 6°. Consequently, the range of average temperatures is much greater in the South. It is about 13° compared with about 8° in the North.

c Probably the South Island because the average winter temperatures are much lower. However, these are averages, and we don't have any information about the range of temperatures from day to day, or during a 24-hour cycle.

2 a Overall their results are not very different. Both do relatively badly in the first test of the year, and then improve.
The IQR for both is 2.

b Brad's results are more consistent. Although their IQRs are the same (2), Abbie's range (10) is greater than Brad's (6).
Brad's results increase after the first week in the term and then remain about the same, whereas Abbie's results increase steadily throughout the term.

c Both Brad's median (15) and mean (14.8) are better than Abbie's (14 and 14.1, respectively).
For the first six weeks of each term he does consistently better. However, for the last four weeks of each term, Abbie gets higher marks than Brad. She would almost certainly beat him in an end-of-year exam.

3 a

	Milk	Juice
IQR	10	14
Range	25	34

b Both show a strong weekly cycle in sales. There are usually fewer of both sold on Mondays, Tuesdays and Wednesdays, and more on Thursdays, with peak sales on Fridays.
The median number of drinks sold each day is similar — 19 for milk and 19.5 for juice.

c The sales of milk follow a much more regular pattern, with the peak sales always on Fridays. These peaks do not fluctuate much, just steadily increase from about 32 to 38 bottles per day.
The sales of juice sometimes peak on Thursday and sometimes on Friday, and their heights vary without any apparent pattern.
The midweek sales of milk are also fairly consistent, but those of juice vary a lot.
Overall the sales of milk appear to be steadily increasing, but those of juice do not.
The mean number of bottles of juice sold (24.16) is higher than that for milk (22.28), so the canteen sells more juice.
Juice sales are much more variable than milk sales. The IQR for juice is 14, whereas it is 10 for milk. The range in the number of bottles of juice sold is 34 and the range for milk is only 25.

d **Argument for milk:**
Sales each day are less variable than for juice, so it will be easier to manage stocks.
Peak sales are always on a Friday, so they will know to prepare for that.
Sales of milk are gradually increasing.
Argument for juice:
They sell more juice — an average of 24.16 bottles per day compared with 22.28 bottles of juice.

e They have only one term's worth of data, and should have more.
We haven't been told from which term the data was collected. Sales may vary between winter and summer, between different sports seasons, etc.

We don't know if there was advertising done for either product, and whether or not it will continue.
We don't know whether the bottles contained similar volumes of drink.
At the end of the year, only Years 9 and 10 are at school, and that might change the pattern of buying drinks.

4 **a** Coffee: $29 000
Ice cream: $12 000–$13 000

b

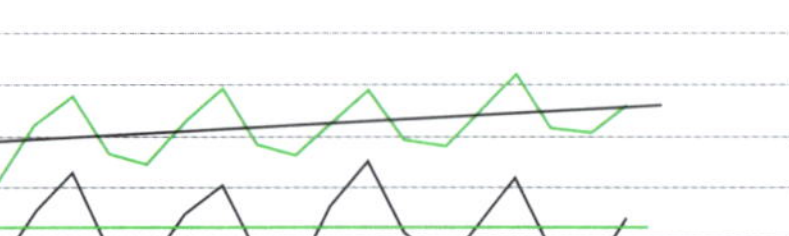

c The shape of the cycles are very similar, with peak sales for both in the March quarter, lowest sales in the September quarter, and second to lowest in the June quarter.

d While the value of ice cream sales doesn't change over all, there is a steady increase in the coffee sales (about $1000 per year). There is also a bigger difference between the peak and lowest quarterly sales for ice cream (about $20 000) than there is for coffee ($15 000 in the first year, $12 000 in the last year).
The difference between lowest and highest quarterly sales does not change for ice creams sales, but it decreases for coffee sales ($15 000 in the first year, $12 000 in the last year).

e

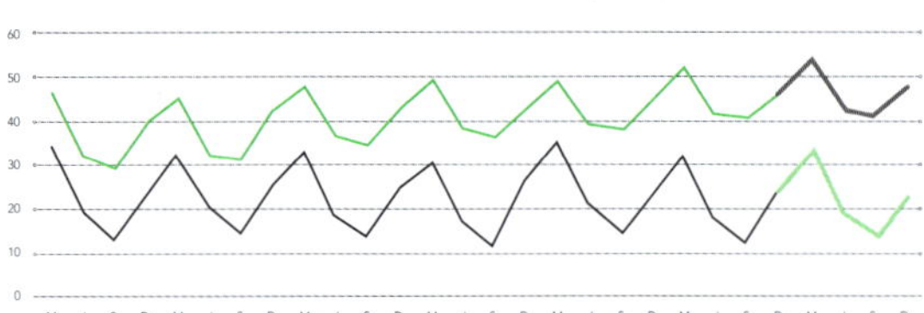

f Reasonably confident because there are six years of fairly consistent data to use as a base. Also, these are monthly averages, so unlikely to be affected by unseasonably hot or cold weather, etc.

Statistical reports (pp. 67–69)

1 **a** In general, it is high (9 000 to 10 000 per month) in December, January and February. It drops through autumn to the lowest values in winter (2 500 – 4 000 per month, and then increase through the spring to summer levels. However, there are three months with unexpectedly high attendance – April, July and October. These correspond with school holidays.

b In March, because there were nearly 9 000 visitors to the zoo, compared with the average of about 6 500. The three months following March also had higher numbers of visitors than average.

2 There has been an increase of about 8 000 visitors in total over the six years. This represents about 11% in six years, which is not dramatic. He makes it look more dramatic by starting the graph at 60 000 instead of 0. Also, if it hadn't been for the gorillas, which probably added about 4 500 to his annual numbers, the total number of visitors in 2015 would probably have been lower than the number in 2012.

3 **a** Sunday: median = about 310
b Monday: range = 415
c Monday has the biggest range (415). However, the range for Sunday (360) is large, but it also has the biggest IQR (220), and this is a better measure of spread.
d Wednesday and Friday
e The tails at the upper end are much longer. Large numbers of people are likely to visit the zoo on public holidays, which are usually on Mondays and Fridays. Therefore there will be just a few unusual points which will make the whiskers long.

4 The manager should not necessarily be concerned, because the mean weight will be low because he included the three babies in his calculation. He should find the mean of just the adult monkeys, or use the median.

5 **a** 2 871
b 10 558
c 46.91%
d 0.2778
e 0.0092
f 0.0623
g NZ – 30.1%
Aus – 14.6%
China – 6.2%
Others – 9.4%
∴ New Zealand

6 **a** Insects, amphibians and reptiles
b Any value between $700 000 and $800 000.

 ISBN: 9780170370424

Practice questions (pp. 70–75)

1 a P(snake) = $\frac{1}{6} = 0.1\dot{6}$

b P(ladder) = $\frac{2}{6} = 0.\dot{3}$

c P(44) = P(2) + P(6) x P(1)

$= \frac{1}{6} + \frac{1}{6} \times \frac{1}{6}$

$= 0.1944$

d Hemi could get there by throwing (2), (4), (1, 1), (1, 3), or (3, 1).
Tara could get there only by throwing (1), (3) or (2, 1).

P(Hemi) = $\frac{1}{6} + \frac{1}{6} + \frac{1}{36} + \frac{1}{36} + \frac{1}{36} = 0.41\dot{6}$

P(Tara) = $\frac{1}{6} + \frac{1}{6} + \frac{1}{36} = 0.36\dot{1}$

∴ Hemi is more likely to land on a ladder. Tara is wrong.

e $\frac{15}{28} = 0.5357$

f $\frac{9}{28} = 0.3214$

g $\frac{9}{15} = 0.6$

h $\frac{3}{28} \times \frac{2}{27} = 0.0079$

i Increase the number of students surveyed.
Take a random sample of students from all levels, not just from one class at one level.
Older students will be more likely to have played the games at some stage, so surveying students from just one level is likely to be biased.
Their class may also be different from other classes at the same level, so its students may be more or less likely to have played the games.

2 a

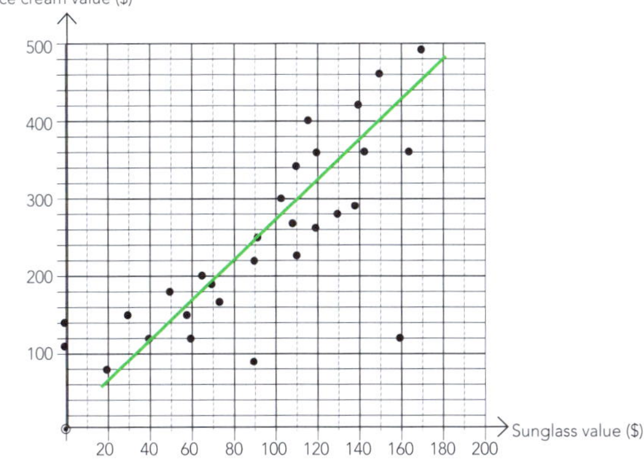

b $30

c 14 days

d (90, 90) and (160, 120). On both days the value of sunglasses sold was relatively high, but the value of ice cream was low. Perhaps these were very bright sunny or snowy days in winter when drivers needed sunglasses, but it was cold so not many wanted ice cream.

e There is a fairly strong positive relationship between sales of sunglasses and ice cream.
The relationship is fairly strong because most points are close to the line of best fit.
The relationship is positive because when the sale of one increases, so does the other. When the weather is hot and sunny, people are likely to buy both, but when it is cold, they are less likely to buy either product.

f 22%

g 25 people

h 23% of 500 = 115

3 a School A: 15.8 (15.7–15.9) in week 20
School B: 16.8 (16.7–16.9) in week 19

b Both classes improved their Algebra throughout each year, and over the two years.
Both classes start in Year 9 at about 8 out of 20.
Both classes reach their overall highest averages in the last two weeks of Year 10.
Both classes reach a peak in weeks 9 or 10 of Year 9 and then their average marks drop to about 9.7 in the first week of Year 10.

c At School A, Algebra marks improve fairly steadily throughout both Year 9 and 10.
At School B, Algebra marks begin at a low level each year, and continue to be low until about halfway through the year, after which they improve very rapidly to reach peaks at the end of each year.
Average marks during the first half of each year are higher at School A, but during the second half of each year they are higher at School B.
The drop in marks at the start of Year 10 was greater at school B.
The average marks at School B are more variable (IQR about 5.7) than School A (IQR about 3).

ISBN: 9780170370424

d School A: looks as though some Algebra is taught or revised throughout both years.
School B: looks as though no algebra is taught or revised for the first half of each year, and but intensively taught during the second half of the year.

e School A did better on average because it had a higher mean (11.82 compared with 11.305) and a higher median (11.35 compared with 10.1).

f Possibly. The median for School B (10.1) lies just outside the box for School A (LQ 10.3), but their boxes overlap. Therefore it is likely that students from school A would do better on average than those from school B.

g If the students are sitting exams at the end of the year, then School B gets to a higher standard by that time. However, at School B the average Algebra marks dropped a lot further between the last test in Year 9 and the first test in Year 10, suggesting that the students forgot more than at School A.
If the students are having assessments during the year, then School A's method of teaching is more effective.

h Only one class was assessed at each school. It would be better to study several classes that have different teachers. Because the data is average marks for the whole class, we do not have information about the spread of marks within each class. Having that information would tell us how consistent the marks are within each class.

ISBN: 9780170370424